BOOK I
CHAPTER IIII – QUBE BYTES:

THE STORY OF OUR LIFE, BASED ON A TRUE LIFE. THE π=3 BOOKS SERIES

*FOR THE RECORD!

JAMES N. AKINS, JR.

The STORY of OUR LIFE,
based on a TRUE LIFE.

The π=3 BOOKS SERIES

The STORY of OUR LIFE, BASED on A TRUE LIFE.
The π = 3 BOOKS SERIES
BOOK 1
CHAPTER 1111 - QUBE BYTES: * FOR THE RECORD!
* THE LIBRARY of CONGRESS CONTROL NUMBER: 2022904632
All Rights Reserved.
Copyright © 2023 James N. Akins, Jr.

Text and Photos - Copyright Registration Number: TX 9-326-438, Registration Decision Date: November 03, 2023

The opinions expressed in this manuscript are solely the opinions of the author and do not represent the opinions or thoughts of the publisher. The author has represented and warranted full ownership and/or legal right to publish all the materials in this book.

This book may not be reproduced, transmitted, or stored in whole or in part by any means, including graphic, electronic, or mechanical without the express written consent of the publisher except in the case of brief quotations embodied in critical articles and reviews.

Paperback ISBN: 979-8-218-96222-7
Hardback ISBN: 979-8-218-96232-6

Cover Art, Illustrations, Unique font selection, Words/phrase meaning of:

"The Story of Our Life, Based on A True Story.
The π = Books Series
** BOOK # - *** Volume #
CHAPTER IIII - QUBE BYES:" /Copyright ©James N. Akins, Jr. 2023.
TITLE
AUTHOR

* FOR REFERENCE ONLY
The actual Library of Congress Control Number: 2022904632 (LOCCN: 2022904632) File is in the Library of Congress and contains all editions of the book: "The Story of Our Life, Based on A True Story." (As well as, original documents featured in the book.). The use of LOCCN: 2022904632 in the title of this unique special International Standard Book Number (ISBN:) book is for reference only. In all future "The π = 3 BOOKS Series" books, the actual "LOCCN: 2022904632 File" in the Library of Congress, represents the physical embodiment of the "lead character" in the series, same as the "Rings" in J.R.R. Tolkiens' "Lord of the Rings" series.
** BOOK #
All English Language books, in the series, shall be BOOK 1, with 1 TITLE, 1 AUTHOR, and unique special International Standard Book Number (ISBN:) per book TITLE.

All Foriegn Language books, in the series, shall be a different BOOK # starting with "BOOK 2- Volume I", with 1 TITLE, 1 AUTHOR, and unique special International Standard Book Number (ISBN:) per book TITLE (i.e. Hebrew Language is BOOK 2 - Volume I, Japaneese Language is BOOK 3 - Volume I, and so on.) Each unique foreign language BOOK # and VOLUME TITLE, AUTHOR and CONTENT shall be written in only the native language of the unique BOOK # and AUTHOR.

*** VOLUME #
After each unique "BOOK # - VOLUME I, TITLE, AUTHOR," 100 % of all previous authors, in a unique "BOOK #" series, shall approve all subsequent unique "VOLUME #" AUTHOR. This is a new, domestic and foreign, book series that is only authorized reproduction and sell shall be in writing from James N. Akins, Jr..

"It has been a long standing understanding the spelling of cube and cubit is correct. In God's Quantum geometry technology, a Qube is equal sides of 1= 9 Qubes (9/9) or 10 Qubebits (!0/10). In God's Quantum Computer technology, there are Qube Bytes. Unlike the present math, the largest whole number is 1, not an infinite number of whole numbers as we are taught now. All this is made clear in my book." /Copyright ©James N. Akins, Jr. 2023

**** © Newton, Isaac, Philosophiae Naturalis Principia Mathematica("Mathematical Principles of Natural Philosophy"), London, 1687; Cambridge, 1713; London, 1726. (Pirated versions of the 1713 edition were also published in Amsterdam in 1714 and 1723*
π = 3 Books
Library of Congress Control Number: 2023915883

PRINTED IN THE UNITED STATES OF AMERICA

The STORY of OUR LIFE, BASED on A TRUE LIFE.
The π = 3 BOOKS SERIES

Table of Content:

Chapter I	BOOK 1, CHAPTER 1111- QUBE BYTES: *FOR THE RECORD! * LIBRARY of CONGRESS CONTROL NUMBER: 2022904632	1
Chapter II	The Story of Our Life, based on a True Life., by Bubba "His X. Mark" Twain	45

I dedicate this book to my Father God, His Son Lord Jesus Christ, and Mother Holy Spirit.

God Bless my Father James Nolan Akins, Sr. and his Mother Cora Black- McMicken and his Father Dr. James E. Burgess (Died 1939) and My Mother Alma Octava Hansen and her Mother Susie Bell Eddins - Hansen (Born 9/16/1908) and her Father Ellis Lee Hansen (Born 4/8/1902) and my sister Alice Fay Akins and my Brother Jack Lee Akins.

God Bless Linda Goodman, my dark haired - brown eyed beutiful Angle in Haiti, that helped God put me back on the right path for me to meet Our Lord Jesus when He returns: Soon! Amend.

.

The STORY of OUR LIFE, BASED on A TRUE LIFE.
The π = 3 BOOKS SERIES

Chapter 1

BOOK 1, CHAPTER 1111 – QUBE BYTES:
*FOR THE RECORD!

*LIBRARY OF CONGRESS CONTROL NUMBER:2022904632

This book is the first book in a series of books, by different authors wishing to explain how God fulfilled His promises to free the world of Satan and the Quantum technology our Lord Jesus Christ will teach all Mankind during His 1000 years reign of peace at our gates.

I have chosen the title of this book: BOOK 1, Library of Congress Control Number: 2022904632, because I have recorded many of the original documents, shown in this book, in the Library of Congress as official historic documents. Like the Holy Bible, this is a written record of God's truths, not a fictitious story book.

Each English speaking author, when approved, will be assigned a unique BOOK 1- Volume Number. 100% of the Book 1 authors must approve all subsequent authors in the BOOK 1 series.

Each Non-English speaking author, when approved, will be assigned a unique BOOK Number- Volume Number. For example, if the first University is Hebrew, it will be BOOK 2, Volume 1. To qualify, the Hebrew University must provide me with a Audio Book quality reading of my book; The Story of Our Life, Based on A True Life. 100% of the Book 2 authors must approve all subsequent authors in the BOOK 2 series. I will have no input on who the authors will be.

It has been a long standing understanding the spelling of cube and cubit is correct. In God's Quantum geometry technology, a Qube is equal sides of 1= 9 Qubes (9/9) or 10 Qubebits (!0/10). In God's Quantum Computer technology, there are Qube Bytes. Unlike the present math, the largest whole number is 1, not an infinite number of whole numbers as we are taught now. All this is made clear in my book.

Receipt 1

SENDER: COMPLETE THIS SECTION

- Complete items 1, 2, and 3.
- Print your name and address on the reverse so that we can return the card to you.
- Attach this card to the back of the mailpiece, or on the front if space permits.

1. Article Addressed to:

Library of Congress
US Programs, Law, and Literature Division
Cataloging in Publication Program
101 Independence Avenue, S.E.
Washington, DC 20540-4283

9590 9402 6713 1060 9641 17

2. Article Number (Transfer from service label)

7020 3160 0002 0446 4009

COMPLETE THIS SECTION ON DELIVERY

A. Signature
X
☐ Agent
☐ Addressee

B. Received by (Printed Name)

C. Date of Delivery

D. Is delivery address different from item 1? ☐ Yes
If YES, enter delivery address below: ☐ No

3. Service Type
☐ Adult Signature
☐ Adult Signature Restricted Delivery
☐ Certified Mail®
☐ Certified Mail Restricted Delivery
☐ Collect on Delivery
☐ Collect on Delivery Restricted Delivery
☐ Insured Mail
☐ Insured Mail Restricted Delivery

☐ Priority Mail Express®
☐ Registered Mail™
☐ Registered Mail Restricted Delivery
☐ Signature Confirmation™
☐ Signature Confirmation Restricted Delivery

PS Form 3811, July 2020 PSN 7530-02-000-9053 — Domestic Return Receipt

---

Receipt 2

SENDER: COMPLETE THIS SECTION

- Complete items 1, 2, and 3.
- Print your name and address on the reverse so that we can return the card to you.
- Attach this card to the back of the mailpiece, or on the front if space permits.

1. Article Addressed to:

Coast Professional Inc.
PO Box 526
Albian, NY 14411

9590 9402 7529 2098 3588 48

2. Article Number (Transfer from service label)

7020 3160 0002 0446 3910

COMPLETE THIS SECTION ON DELIVERY

A. Signature
X *Allyn Halesky* ☐ Agent ☐ Addressee

B. Received by (Printed Name)
Allyson Halesky

C. Date of Delivery

D. Is delivery address different from item 1? ☑ Yes
If YES, enter delivery address below: ☐ No

PO Box 425
Geneseo, NY 14454

3. Service Type
☐ Adult Signature
☐ Adult Signature Restricted Delivery
☐ Certified Mail®
☐ Certified Mail Restricted Delivery
☐ Collect on Delivery
☐ Collect on Delivery Restricted Delivery
☐ Insured Mail
☐ Insured Mail Restricted Delivery (over $500)

☐ Priority Mail Express®
☐ Registered Mail™
☐ Registered Mail Restricted Delivery
☐ Signature Confirmation™
☐ Signature Confirmation Restricted Delivery

PS Form 3811, July 2020 PSN 7530-02-000-9053 — Domestic Return Receipt

2021 Third Continental Congress
Carpenters' Hall
Philadelphia, Pennsylvania
March 8, 2023

U. S. Department of the Treasury
National Payment Integrity and Resolution Center
P. O. Box 51315
Philadelphia, Pennsylvania 19115-6314

Greetings U S Department of Treasury,

1. After March 4, 2023, the first order of business of the Third Continental Congress (TCC) is to assure each State National Citizen (SNC) of the 1776 Declaration of Independence and 1787 United States of America Constitution financial and economic retribution for the money Extorted from the SNC (a.k.a "American Tax Payers") since 1915 by the following Unconstitutional and illegal organizations and individuals (Defendants):

Violates Article I, Section 8, Clause 1 and/or Article XI of the 1787 United States of America Constitution.

1. Internal Revenue Service (IRS),
2. Bankrupted United States of America, INC. (USAINC),
3. Bankrupted United Nations, INC. (UNINC),
4. UNINC US Office of Personnel Management (OPM) (i.e. FBI, CIA, Congress, IRS, President, Supreme Court Justices, etc.)
5. British Accreditation Registry-Crown Temple British Maritain Law (B.A.R.)
6. The Central Bank of the USAINC is the Private Federal Reserve System, created by Congress in 1913.
7. Private Western Central Vatican Bank and Central Bank of London.
8. North Atlantic Treaty Organization (NATO)

2. All State National Citizens (a.k.a. American Tax Payers) shall not have any direct contact with any of the "Defendants" stated above. If one is approached by any individual claiming to representing the above stated "Defendants", do not be adversarial or aggressive, be polite and helpful. Report all encounters and all relevant correspondences (Postal Mail, E-Mail, Phone and personal contact reports) to the local Constitutional County Sheriffs' office.

3. The Third Continental Congress Attorney General shall coordinate with the Governor of each of the 50 United States during the resolution of this Extortion Case and provide a progress report weekly.

4. In compliance with paragraph 2 above, I wish to report Exhibit II, from the Department of the Treasury, Internal Revenue Service, Philadelphia, Pennsylvania 19255, in my case.

Praise God for His gifts of Life, Liberty, and the Pursuit of Happiness to all mankind.

James N. Akins, Jr. © Without Prejudice, P. O. Box 1114, Fairhope, Alabama 36533-1111

Cc: Department of the Treasury, Program Manager, Mary Salnaitis Internal Revenue Service, Philadelphia, Pennsylvania 19255

Department of the Treasury
Internal Revenue Service

PHILADELPHIA PA 19255

In reply refer to: 0544300093
Mar. 07, 2023   LTR 2645C   K0
***-**-5199   202212 30
Input Op: 0509906021 00027398
BODC: WI

JAMES N AKINS JR
PO BOX 1111
FAIRHOPE   AL   36533-1111

021060

Taxpayer identification number: ***-**-5199
Tax periods: Dec. 31, 2022

Form: 1040

Dear Taxpayer:

Thank you for your inquiry of Jan. 24, 2023.

We're working on your account. However, we need an additional 60 days to send you a complete response on what action we are taking on your account. We don't need any further information from you right now.

If you prefer, you can write to that office at the address we provided in this letter.

If you have questions, you can call 1-800-829-0922.

If you prefer, you can write to the address at the top of the first page of this letter.

Find tax forms or publications by visiting www.irs.gov/forms or calling 800-TAX-FORM (800-829-3676).

Whenever you write, include a copy of this letter and your telephone numbers along with the hours we can reach you.

Keep a copy of this letter for your records.

Thank you for your cooperation.

EXHIBIT II

```
                                        0544300093
                              Mar. 07, 2023  LTR 2645C  K0
                              ***-**-5199   202212 30
                              Input Op:  0509906021 00027399
```

JAMES N AKINS JR
PO BOX 1111
FAIRHOPE   AL   36533-1111

Sincerely yours,

Mary Salnaitis

Mary Salnaitis
Program Manager

2021 Third Continental Congress
Carpenters' Hall
Philadelphia, Pennsylvania
March 13, 2023

U. S. Department of the Treasury
National Payment Integrity and Resolution Center
P. O. Box 51315
Philadelphia, Pennsylvania 19115-6314

Greetings U S Department of Treasury,

1. After March 4, 2023, the first order of business of the Third Continental Congress (TCC) is to assure each State National Citizen (SNC) of the 1776 Declaration of Independence and 1787 United States of America Constitution financial and economic retribution for the money Extorted from the SNC (a.k.a "American Tax Payers") since 1915 by the following Unconstitutional and illegal organizations and individuals (Defendants):

Violates Article I, Section 8, Clause 1 and/or Article XI of the 1787 United States of America Constitution.

1. Internal Revenue Service (IRS),
2. Bankrupted United States of America, INC. (USAINC),
3. Bankrupted United Nations, INC. (UNINC),
4. UNINC US Office of Personnel Management (OPM) (i.e. FBI, CIA, Congress, IRS, President, Supreme Court Justices, etc.)
5. British Accreditation Registry-Crown Temple British Maritain Law (B.A.R.)
6. The Central Bank of the USAINC is the Private Federal Reserve System, created by Congress in 1913.
7. Private Western Central Vatican Bank and Central Bank of London.
8. North Atlantic Treaty Organization (NATO)

2. All State National Citizens (a.k.a. American Tax Payers) shall not have any direct contact with any of the "Defendants" stated above. If one is approached by any individual claiming to representing the above stated "Defendants", do not be adversarial or aggressive, be polite and helpful. Report all encounters and all relevant correspondences (Postal Mail, E-Mail, Phone and personal contact reports) to the local Constitutional County Sheriffs' office.

3. The Third Continental Congress Attorney General shall coordinate with the Governor of each of the 50 United States during the resolution of this Extortion Case and provide a progress report weekly.

4. In compliance with paragraph 2 above, I wish to report Exhibit III, from the Department of the Treasury, Internal Revenue Service, Philadelphia, Pennsylvania 19255, in my case.

Praise God for His gifts of Life, Liberty, and the Pursuit of Happiness to all mankind.

[Signature]

James N. Akins, Jr. © Without Prejudice, P. O. Box 1111, Fairhope, Alabama 36533-1111

Cc: Department of the Treasury, Program Manager, Mary Salnaitis Internal Revenue Service, Philadelphia, Pennsylvania 19255

Department of the Treasury
Internal Revenue Service

PHILADELPHIA PA 19255

In reply refer to: 0547595410
Mar. 03, 2023   LTR 86C   0
***-**-5199   202212 30
                00000218
BODC: SB

JAMES N AKINS JR
PO BOX 1111
FAIRHOPE AL 36533-1111

Tax periods: Dec. 31, 2022

Dec. 31, 2022

Dear Taxpayer:

Thank you for your correspondence of Jan. 20, 2023.

We're sending your claim, Form 1040, to the Ogden Customer Service Center to process. That office will contact you within 60 days.

Find tax forms or publications by visiting www.irs.gov/forms or calling 800-TAX-FORM (800-829-3676).

If you have questions, you can call 800-829-0922.

If you prefer, you can write to the address at the top of the first page of this letter.

When you write, include a copy of this letter, and provide your telephone number and the hours we can reach you in the spaces below.

Telephone number ( )_____ Hours _____

Keep a copy of this letter for your records.

Thank you for your cooperation.

```
                                              0547595410
                          Mar. 03, 2023    LTR 86C    0
                          ***-**-5199    202212 30
                                                00000219
```

JAMES N AKINS JR
PO BOX 1111
FAIRHOPE   AL   36533-1111

                              Sincerely yours,

                              Melissa Franklin
                              Melissa Franklin
                              Operation 2 Manager

Enclosures:
Envelope

Department of the Treasury
Internal Revenue Service

PHILADELPHIA PA 19255

000933.501323.345870.8023 1 MB 0.531 702

JAMES N AKINS JR
PO BOX 1111
FAIRHOPE AL 36533-1111

CUT OUT AND RETURN THE VOUCHER IMMEDIATELY BELOW IF YOU ONLY HAVE AN INQUIRY.
DO NOT USE IF YOU ARE MAKING A PAYMENT.

CUT OUT AND RETURN THE VOUCHER AT THE BOTTOM OF THIS PAGE IF YOU ARE MAKING A PAYMENT,
EVEN IF YOU ALSO HAVE AN INQUIRY.

The IRS address must appear in the window.
0547595410

BODCD-

Use for inquiries only	
Letter Number:	LTR0086C
Letter Date :	2023-03-03
Tax Period :	202212

*****5199

JAMES N AKINS JR
PO BOX 1111
FAIRHOPE AL 36533-1111

INTERNAL REVENUE SERVICE

PHILADELPHIA PA 19255

421765199 RC AKIN 30 0 202212 670 00000000000

The IRS address must appear in the window.
0547595410

BODCD-

Use for payments	
Letter Number:	LTR0086C
Letter Date :	2023-03-03
Tax Period :	202212

*****5199

JAMES N AKINS JR
PO BOX 1111
FAIRHOPE AL 36533-1111

INTERNAL REVENUE SERVICE

KANSAS CITY MO 64999-0202

421765199 RC AKIN 30 0 202212 670 00000000000

URGENT: PLEASE FILE BEFORE YOUR BANK CLOSES.

As of March 4, 2023, God fulfilled Genesis 6:3 and Daniel 2:35. We The People of the Second Constitutional Republic of The United States of America celebrate the start of 121st Jubilee- The beginning of Christ's Reign, by declaring all the wealth stolen by the unconstitutional Internal Revenue Service (IRS) and Federal Reserve Bank.

All American taxpayers and their family members estimate all the wealth that was stolen from them since 1915, and multiply by 7 (Proverbs 6:31, Example $1,000,000 x 7 = $7,000,000 + $1,000,000 = $8,000,000). No documentation required at any time.

All American taxpayers must have the opportunity to put in their claim before a resolution may be issued.

First Step: Declare your intent to receive a resolution. On the front of 2 envelope or 54 cent USP Post Card, put your name and address, as it appears on your IRS Tax Forms, as the RETURN ADDRESS; ADDRESS TO: U.S. Department of the Treasury (#1 Envelope or Post Card) and Department of the Treasury (Envelope #2 or Post Card); on the back of each Envelope or 54 cent USP Post Card write, " I estimate my stolen wealth at a minimum of $X,XXX,XXX.XX. Genesis 6:3, Proverbs 6:31 and Daniel 2:35" DO NOT SEND ANY DOCUMENTS OR WRITE SOCIAL SECURITY NUMBERS.

1. U.S. Department of the Treasury
   National Payment Integrity and Resolution Center
   P.O Box 51315
   Philadelphia, Pennsylvania 19115-6314

2. Department of the Treasury
   Program Manager, Mary Salnaitis
   Internal Revenue Service
   Philadelphia, Pennsylvania 19255

Second Step: Download FREE ebook, "The Story of Our Life, Based on A True Life." Library of Congress Control Number:2022904632. The ebook is available to anyone in the world by accessing www.piisthree,com or www.piis3.com , scrolling down to "Our Products" , click on "Free ebook" and Download.

After Downloading the Free ebook; print pages 114, 115 and 116: estimate all money stolen from you and/or your ancestors back to 1915 and multiply it by 10 for fees and interest (No documentation required for estimating.); Complete altered 1040 form and follow mailing instructions on page 116.

James N. Akins, Jr. © without prejudice
P. O. Box 1111
Fairhope, AL 36533-1111

U.S. Department of the Treasury
National Payment Integrity and
Resolution Center
P. O Box 51315
Philadelphia, Pennsylvania 19115-6314

I estimate my stolen wealth at a
minimum of $ 10,000,000.00
Genesis 6:3, Proverbs 6:31 and Daniel 2:35

Ten Million Dollars

I estimate my stolen wealth at a minimum of $ 10,000,000.00
Genesis 6:3, Proverbs 6:31 and Daniel 2:35

Ten Million Dollars

James N. Akins, Jr. © without prejudice
P. O. Box 1111
Fairhope, AL 36533-1111

Department of the Treasury
Program Manager, Mary Salnaitis
Internal Revenue Service
Philadelphia, Pennsylvania 19255

2021 Third Continental Congress
Carpenters' Hall
Philadelphia, Pennsylvania
March 30, 2023

U. S. Department of the Treasury
National Payment Integrity and Resolution Center
P. O. Box 51315
Philadelphia, Pennsylvania 19115-6314

Greetings U S Department of Treasury,

1. After March 4, 2023, the first order of business of the Third Continental Congress (TCC) is to assure each State National Citizen (SNC) of the 1776 Declaration of Independence and 1787 United States of America Constitution financial and economic retribution for the money Extorted from the SNC (a.k.a "American Tax Payers") since 1915 by the following Unconstitutional and illegal organizations and individuals (Defendants):

Violates Article I, Section 8, Clause 1 and/or Article XI of the 1787 United States of America Constitution.

1. Internal Revenue Service (IRS),
2. Bankrupted United States of America, INC. (USAINC),
3. Bankrupted United Nations, INC. (UNINC),
4. UNINC US Office of Personnel Management (OPM) (i.e. FBI, CIA, Congress, IRS, President, Supreme Court Justices, etc.)
5. British Accreditation Registry-Crown Temple British Maritain Law (B.A.R.)
6. The Central Bank of the USAINC is the Private Federal Reserve System, created by Congress in 1913.
7. Private Western Central Vatican Bank and Central Bank of London.
8. North Atlantic Treaty Organization (NATO)

2. All State National Citizens (a.k.a. American Tax Payers) shall not have any direct contact with any of the "Defendants" stated above. If one is approached by any individual claiming to representing the above stated "Defendants", do not be adversarial or aggressive, be polite and helpful. Report all encounters and all relevant correspondences (Postal Mail, E-Mail, Phone and personal contact reports) to the local Constitutional County Sheriffs' office.

3. The Third Continental Congress Attorney General shall coordinate with the Governor of each of the 50 United States during the resolution of this Extortion Case and provide a progress report weekly.

4. In compliance with paragraph 2 above, I wish to report Exhibit IIII, from the Department of the Treasury, Internal Revenue Service, Philadelphia, Pennsylvania 19255, in my case.

Praise God for His gifts of Life, Liberty, and the Pursuit of Happiness to all mankind.

[signature]
James N. Akins, Jr. © Without Prejudice P. O. Box 1111, Fairhope, Alabama 36533-1111

Cc: Department of the Treasury, Program Manager, Mary Salnaitis Internal Revenue Service, Philadelphia, Pennsylvania 19255

Department of the Treasury, Director Return Integrity Verification Ops., Gardy Larochelle, Internal Revenue Service, 3651 S IH 35, STOP 6579 AUSC, Austin, Texas 73301-0059

IRS Department of the Treasury
Internal Revenue Service

3651 S IH 35, STOP 6579 AUSC
AUSTIN TX 73301-0059

In reply refer to: 1485011111
Mar. 27, 2023 LTR 5071C B0
***-**-5199 202112 30
                00225112
              BODC: WI

JAMES N AKINS JR
PO BOX 1111
FAIRHOPE AL 36533

```
           Tax year: 2021
           Tax form: 1040SR
     Control number: 29205035118043
      Letter number: 5071C
```

Dear TAXPAYER

WHY WE'RE SENDING THIS LETTER

We received a tax return for tax year 2021 using your name and Social Security number (SSN) or individual taxpayer identification number (ITIN).

If you DID file: Verify your return online at IRS.gov/VerifyReturn. We need more information from you to verify your identity and tax return information so we can continue processing your tax return and issue a refund, or credit any overpayments to your next year's estimated tax. We can't process your tax return until we hear from you.

If you DIDN'T file: Report this at IRS.gov/VerifyReturn immediately to confirm that you did not file a tax return, as you may be a victim of identity theft.

When you contact us, you must refer to control number: 29205035118043.

DÓNDE PUEDE OBTENER INFORMACIÓN EN ESPAÑOL

Usted puede solicitar una copia de esta carta en español, llamando al número de teléfono 800-830-5084 entre 7:00 a.m. y 7:00 p.m. horario local opcion 2.

Para obtener más información sobre esta carta, visite IRS.gov/ltr5071sp.

WHAT YOU NEED TO DO

Most taxpayers in your situation successfully verified their identity and tax return online within a few days, enabling us to process their tax returns immediately.

You must verify your return at IRS.gov/VerifyReturn. You'll be asked to either create an account and verify your identity or sign in with

EXHIBIT IIII

```
                                                      1485011111
                                    Mar. 27, 2023  LTR 5071C  B0
                                    ***-**-5199   202112  30
                                                         00225113
```

JAMES N AKINS JR
PO BOX 1111
FAIRHOPE AL 36533

an existing account. After signing in, you'll then be asked questions about your tax return.

IMPORTANT: We won't be able to process your tax return until you answer the tax return questions. If you verified your identity, but didn't answer the tax return questions, you must sign in again at IRS.gov/VerifyReturn and answer the questions.

To complete the verify your return questions, you'll need:
- This letter
- The Form 1040-series tax return for the tax year shown above (Forms W-2 and 1099 aren't tax returns)

OTHER WAYS TO VERIFY YOUR TAX RETURN

If you can't use our Verify Your Return online, you can call our Taxpayer Protection Program hotline at 800-830-5084 between 7:00 a.m. and 7:00 p.m., local time.

When you call, you MUST have ALL of the following:
- This letter
- The Form 1040-series tax return for the tax year shown above (Forms W-2 and 1099 aren't tax returns)
- A prior year tax return other than the year shown above, if you filed one (Forms W-2 and 1099 aren't tax returns)
- Supporting documents for each year's tax return you filed (e.g., Form W-2, Form 1099, Schedule C, Schedule F, etc.)

WHAT WE'LL DO AFTER YOU VERIFY YOUR TAX RETURN

If you've successfully verified your return, it may take up to 9 weeks for you to receive your refund or apply a credit of any overpayment to your account. However, if there are other issues, you may receive a notice asking for more information, and this may delay your refund.

If we can't verify your identity online or over the phone, we'll ask you to schedule an appointment and bring the documents listed above to your local IRS office to verify in person.

WHERE YOU CAN GO FOR MORE INFORMATION

Visit IRS.gov/ltr5071c for information about this letter.

Keep a copy of this letter for your records.

Thank you for your cooperation.

```
                                            1485011111
                            Mar. 27, 2023  LTR 5071C  B0
                            ***-**-5199   202112 30
                                               00225114
```

JAMES N AKINS JR
PO BOX 1111
FAIRHOPE   AL   36533

Sincerely yours,

Gardy Larochelle

Gardy Larochelle, Director
Return Integrity Verification Ops.

April 6, 2023

Greetings to all 1787 United States Constitution County Sheriffs:

1. My name is James N. Akins, Jr. a property owner/resident at 18667 N Greeno Road, Fairhope, Alabama 36532, Baldwin County. I wish to report mail fraud being perpetrated on We the People of the 1787 United States Constitution. I am the first claimant in the Second Constitutional Republic of The United States of America (SCRUSA) Attorney General's indictment of the below listed Unconstitutional and illegal organizations and individuals (Defendants) each in violation of the 1776 Declaration of Independence and 1787 United States of America Constitution.

2. I have chronicled all correspondence in: "The Story of Our Life, Based on A True Life", Library of Congress Control Number: 2022904632. Free ebook copies at www.piisthree.com or www.Piis3.com under "Our Products". The bankrupt defunct Internal Revenue Service (IRS), et al listed below, all closed on December 21, 2022. Therefore now, these Corporations do not have a physical address in order to receive mails legally. These mails need to be delivered to the physical street address in the state where one's company was formed. Thus, one needs to have a registered agent who receives your official mails in case you have an LLC or a corporation.

3. On December 22, 2022, the Second Constitutional Republic of The United States of America was established according to the 1787 United States Constitution. On March 4, 2023 and in compliance with 1787 United States Constitution, Article II, Section 2, Clause 1, Donald John Trump was inaugurated the first president and Commander-in-Chief of the 1787 United States Constitutional Military of the Second Constitutional Republic of The United States of America.

4. To assure financial and economic retribution for the wealth Extorted from the State National Citizen (SNC) (a.k.a "American Tax Payers"), since 1915, the Second Constitutional Republic of The United States of America (SCRUSA) Attorney General is indicting the following for violationing Article I, Section 8, Clause 1 and/or Article XI of the 1787 United States of America Constitution.

1. Internal Revenue Service (IRS),
2. Bankrupted United States of America, INC. (USAINC),
3. Bankrupted United Nations, INC. (UNINC),
4. UNINC US Office of Personnel Management (OPM) (i.e. FBI, CIA, Congress, IRS, President, Supreme Court Justices, etc.)
5. British Accreditation Registry-Crown Temple British Maritain Law (B.A.R.)
6. The Central Bank of the USAINC is the Private Federal Reserve System, created by Congress in 1913.
7. Private Western Central Vatican Bank and Central Bank of London.
8. North Atlantic Treaty Organization (NATO)

5. All State National Citizens (a.k.a. American Tax Payers) shall not have any direct contact with any of the "Defendants" stated above. If one is approached by any individual claiming to representing the above stated "Defendants", do not be adversarial or aggressive, be polite and helpful. Report all encounters and all relevant correspondences (Postal Mail, E-Mail, Phone and personal contact reports) to the local Constitutional County Sheriffs' office. There shall not be any mail correspondence between any of the defunct corporations listed above. Do not open

any mail from the Department of the Treasury, IRS and RETURN TO SENDER to prove you did not contract with them.

6. As of March 4, 2023, only send and receive correspondence with U. S. Department of the Treasury
National Payment Integrity and Resolution Center, P. O. Box 51315, Philadelphia, Pennsylvania 19115-6314. As soon as possible, 1. Send the US Department of Treasury wealth a post card with your IRS return address on the front and write the following on the back; " I estimate my stolen wealth at a minimum of $_____(Genesis 6:3, Proverbs 6:31, and Daniel 2:35)". 2. Download and print pages 114, 115, and 116 in the ebook, complete altered IRS 1040-SR form, mail to ONLY the US Department of the Treasury. NO ADDITIONAL BACKUP DOCUMENTATION WILL BE REQUIRED. Correspondence between myself and the Department of the Treasury, IRS, is in my ebook to prove the IRS has acknowledged the ongoing SCRUSA Attorney General indictment.

7. The Second Constitutional Republic` of The United States of America's Attorney General shall coordinate with the Governor of each of the 50 United States during the resolution of this Extortion Case and provide a progress report weekly.

Praise God for His gifts of Life, Liberty, and the Pursuit of Happiness to all mankind.

[signature: James N Akins, Jr.]

James N. Akins, Jr. © Without Prejudice, P. O. Box 1111, Fairhope, Alabama 36533-1111

Cc:

U. S. Department of the Treasury
National Payment Integrity and Resolution Center
P. O. Box 51315
Philadelphia, Pennsylvania 19115-6314

Definitions:
___

Defunct, in business context, refers to the condition of a of a company, wheither publicly yarded or private, that has gone bankrupt and has ceased to exist. Typically, "defunct" that is no longer existing, functioning, or in use.

Now, to register an LLC, you need to have a physical address in order to receive mails legally. These mails need to be delivered to the physical street address in the state where your company was formed. Thus, you need to have a registered agent who receives your official mails in case you have an LLC or a corporation.

*To register, you will need a USPS **Business** Customer Gateway account, a PostalOne! permit, a USPS Enterprise Payment System (EPS) account, an Electronic Verification System (eVS ®) account, and a valid **business** address. Simply pay the annual enrollment fee and complete the online application. Register Now Increase Your Own Mail Productivity*

From: **James Akins** mr.akins21@yahoo.com
Subject: **Fwd: TCC Oath of Office**
Date: April 10, 2023 at 5:56 AM
To: **James Akins** Mr.Akins21@yahoo.com

Begin forwarded message:

From: James Akins <mr.akins21@yahoo.com>
Subject: Fwd: TCC Oath of Office
Date: March 26, 2023 at 10:08:16 PM PDT
To: john white <jcw652002@yahoo.com>

John,

You are appointed as Speaker of the TCC House until all 50 states ratify the updated us constitution. The We the People of the 2nd Con. Rep. of the USA have elections of the 200 reps and Speaker of the house. The TCC will be adjourned.

Please, on 2 copies, Sign and write YES or NO at the end. Keep a copy, scan and email me. Mail a copy to:

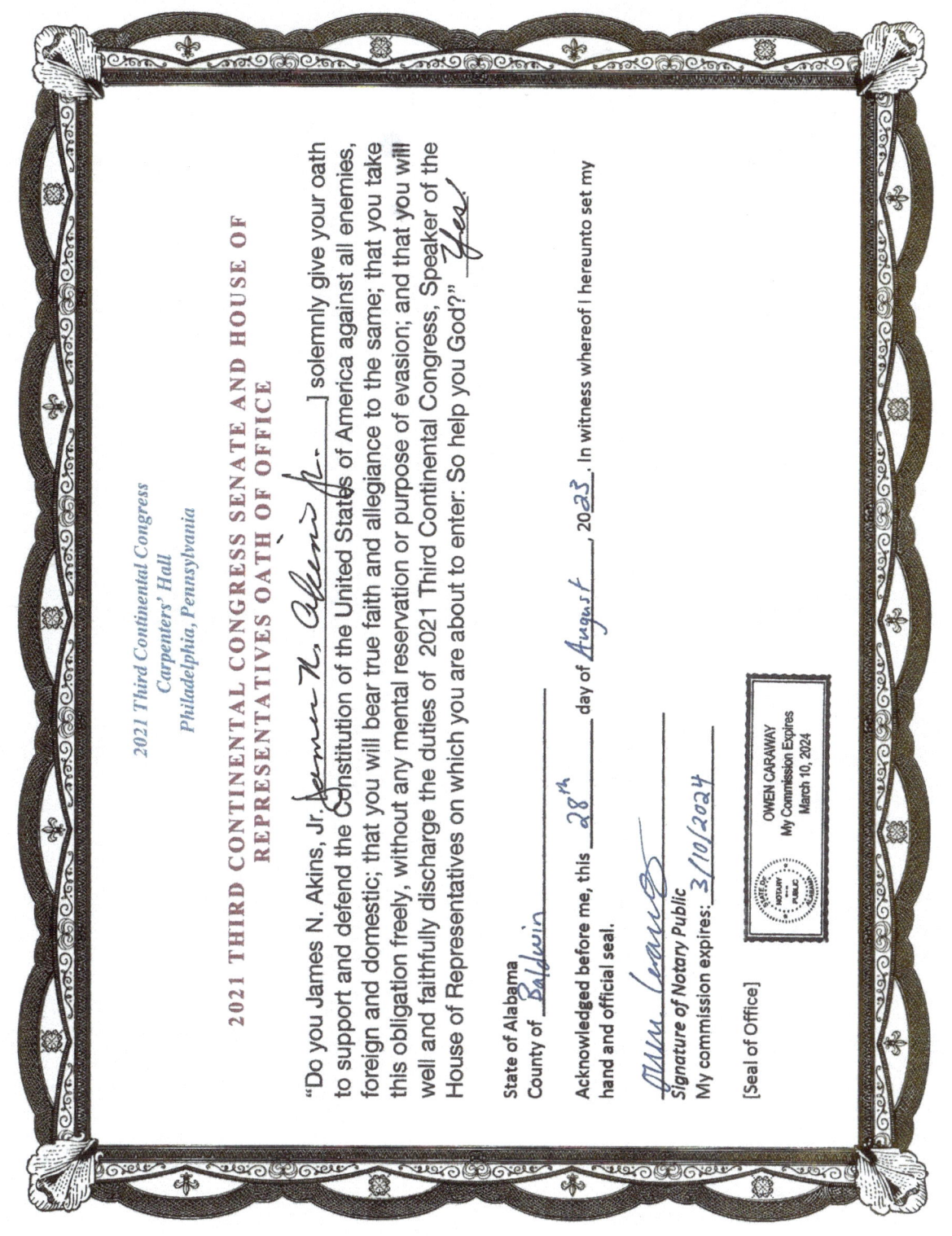

We the People of the Great State of California
California Governor's Office
State Capitol
1st Fl.
Sacramento, CA 95814

U.S. Department of the Treasury
National Payment Integrity and
Resolution Center
P. O Box 51315
Philadelphia, Pennsylvania
19115-6314

Second Constitutional Republic of
The United States of America
&
2021 Third Continental Congress
P. O. Box 1111
Fairhope, Alabama 36533-1111

I estimate our stolen wealth at a minimum of $ 14.1 Trillion.
Genesis 6:3, Proverbs 6:31 and Daniel 2:35

Second Constitutional Republic of
The United States of America
&
2021 Third Continental Congress
April 7, 2023
WWG1WGA

We the People of the Great State of Alabama
Alaska Governor's Office
State Capitol
P.O. Box 110001
Juneau, AK 99811

Greetings Constitutional County Sheriffs of the Great State of Alabama:

1. My name is James N. Akins, Jr. a property owner/resident at 18667 N Greeno Road, Fairhope, Alabama 36532, Baldwin County Alabama. I wish to report mail fraud being perpetrated on We the People of the 1787 United States Constitution. I am the first claimant in the Second Constitutional Republic` of The United States of America (SCRUSA) Attorney General's indictment of the below listed Unconstitutional and illegal organizations and individuals (Defendants):each of the 1776 Declaration of Independence and 1787 United States of America Constitution.

2. I have chronicled all correspondence in: "The Story of Our Life, Based on A True Life", Library of Congress Control Number: 2022904632. Free ebook copies at www.piisthree or www.Piis3.com under "Our Products". The bankrupt defunct Internal Revenue Service (IRS), et al listed below, all closed on December 21, 2022. Therefore now, these Corporations do not have a physical address in order to receive mails legally. These mails need to be delivered to the physical street address in the state where one's company was formed. Thus, one needs to have a registered agent who receives your official mails in case you have an LLC or a corporation.

3. On December 22, 2022, the Second Constitutional Republic` of The United States of America was established according to the 1787 United States Constitution. On March 4, 2023 and in compliance with 1787 United States Constitution, Article II, Section 2, Clause 1, Donald John Trump was inaugurated the first president and Commander-in-Chief of the 1787 United States Constitutional Military of the Second Constitutional Republic` of The United States of America.

4. To assure financial and economic retribution for the wealth Extorted from the State National Citizen (SNC) (a.k.a "American Tax Payers"), since 1915, the Second Constitutional Republic` of The United States of America (SCRUSA) Attorney General is indicting the following for violates **Article I, Section 8, Clause 1 and/or** Article XI of the 1787 **United States of America Constitution.**

   1. Internal Revenue Service (IRS),
   2. Bankrupted United States of America, INC. (USAINC),
   3. Bankrupted United Nations, INC. (UNINC),
   4. UNINC US Office of Personnel Management (OPM) (i.e. FBI, CIA, Congress, IRS, President, Supreme Court Justices, etc.)
   5. Bitish Accreditation Registry-Crown Temple British Maritain Law (B.A.R.)
   6. The Central Bank of the USAINC is the Private Federal Reserve System, created by Congress in 1913.
   7. Private Western Central Vatican Bank and Central Bank of London.
   8. North Atlantic Treaty Organization (NATO)

5. All State National Citizens (a.k.a. American Tax Payers) shall not have any direct contact with any of the "Defendants" stated above. If one is approached by any individual claiming to representing the above stated "Defendants", do not be adversarial or aggressive, be polite and helpful. Report all encounters and all relevant correspondences (Postal Mail, E-Mail, Phone and personal contact reports) to the local Constitutional County Sheriffs' office. There shall not be any mail correspondence between any of the defunct corporations listed above. Do not open any mail from the Department of the Treasury, IRS and RETURN TO SENDER to prove you did not contract with them.

6. As of March 4, 2023, only send and receive correspondence with U. S. Department of the Treasury National Payment Integrity and Resolution Center, P. O. Box 51315, Philadelphia, Pennsylvania 19115-6314. As soon as possible, 1. Send the US Department of Treasury wealth a post card with your IRS return address on the front and write the following on the back; " I estimate my stolen wealth at a minimum of $_____(Genesis 6:3, Proverbs 6:31, and Daniel 2:35)". 2. Download and print pages 114, 115, and 116 in the ebook, complete altered IRS 1040-SR form, mail to ONLY the US Department of the Treasury. NO ADDITIONAL BACKUP DOCUMENTATION WILL BE REQUIRED. Correspondence between myself and the Department of the Treasury, IRS, is in my ebook to prove the IRS has acknowledged the ongoing SCRUSA Attorney General indictment.

7. The Second Constitutional Republic` of The United States of America's Attorney General shall coordinate with the Governor of each of the 50 United States during the resolution of this Extortion Case and provide a progress report weekly.

Praise God for His gifts of Life, Liberty, and the Pursuit of Happiness to all mankind.

[signature]

James N. Akins, Jr. © Without Prejudice, P. O. Box 1111, Fairhope, Alabama 36533-1111

Cc:

U. S. Department of the Treasury
National Payment Integrity and Resolution Center
P. O. Box 51315
Philadelphia, Pennsylvania 19115-6314

Definitions:

Defunct, in business context, refers to the condition of a company, whether publicly yarded or private, that has gone bankrupt and has ceased to exist. Typically, "defunct" that is no longer existing, functioning, or in use.

Now, to register an LLC, you need to have a physical address in order to receive mails legally. These mails need to be delivered to the physical street address in the state where your company was formed. Thus, you need to have a registered agent who receives your official mails in case you have an LLC or a corporation.

*To register, you will need a USPS **Business** Customer Gateway account, a PostalOne! permit, a USPS Enterprise Payment System (EPS) account, an Electronic Verification System (eVS ®) account, and a valid **business** address. Simply pay the annual enrollment fee and complete the online application. Register Now Increase Your Own Mail Productivity.*

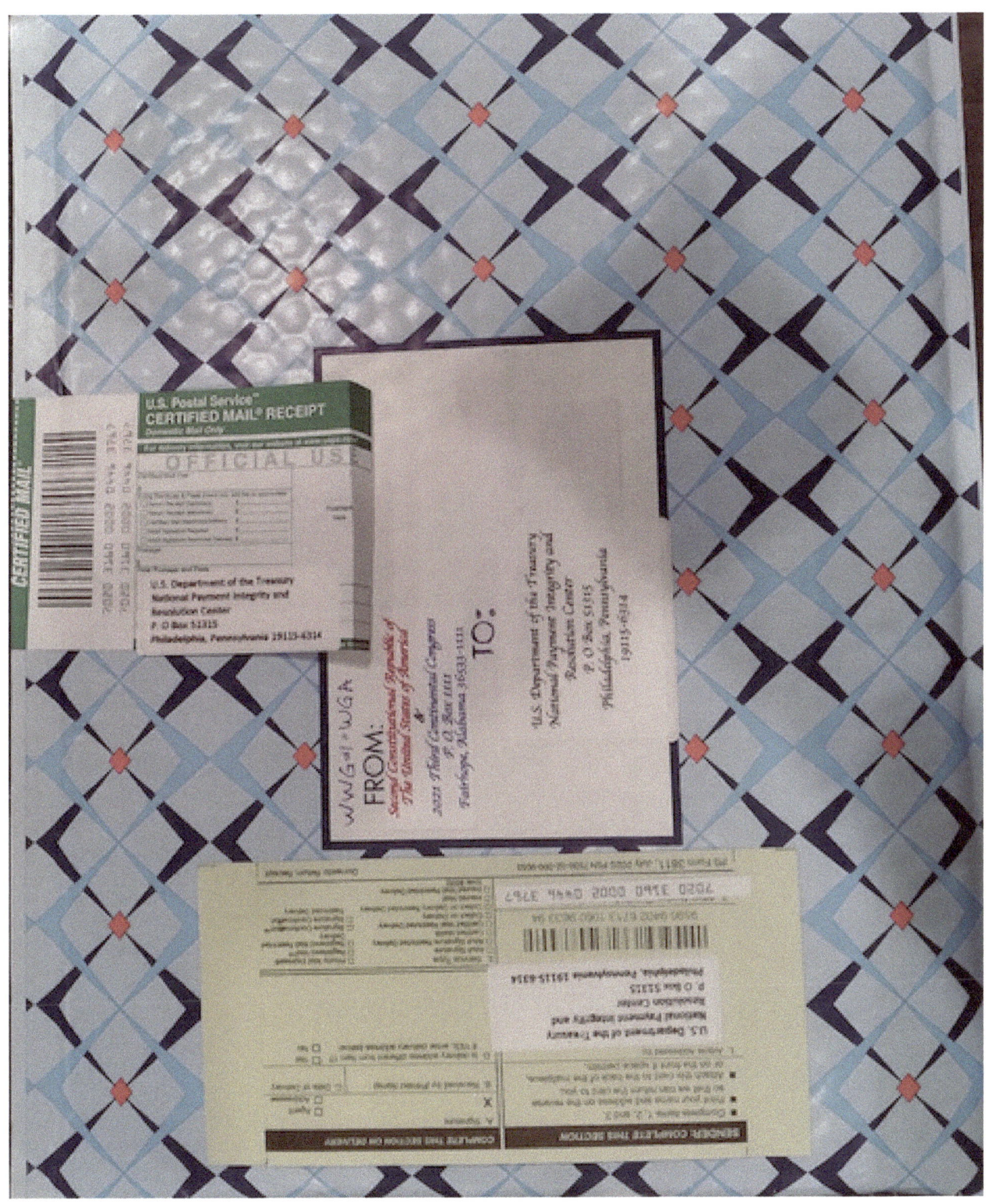

Mailed on April 10, 2023.

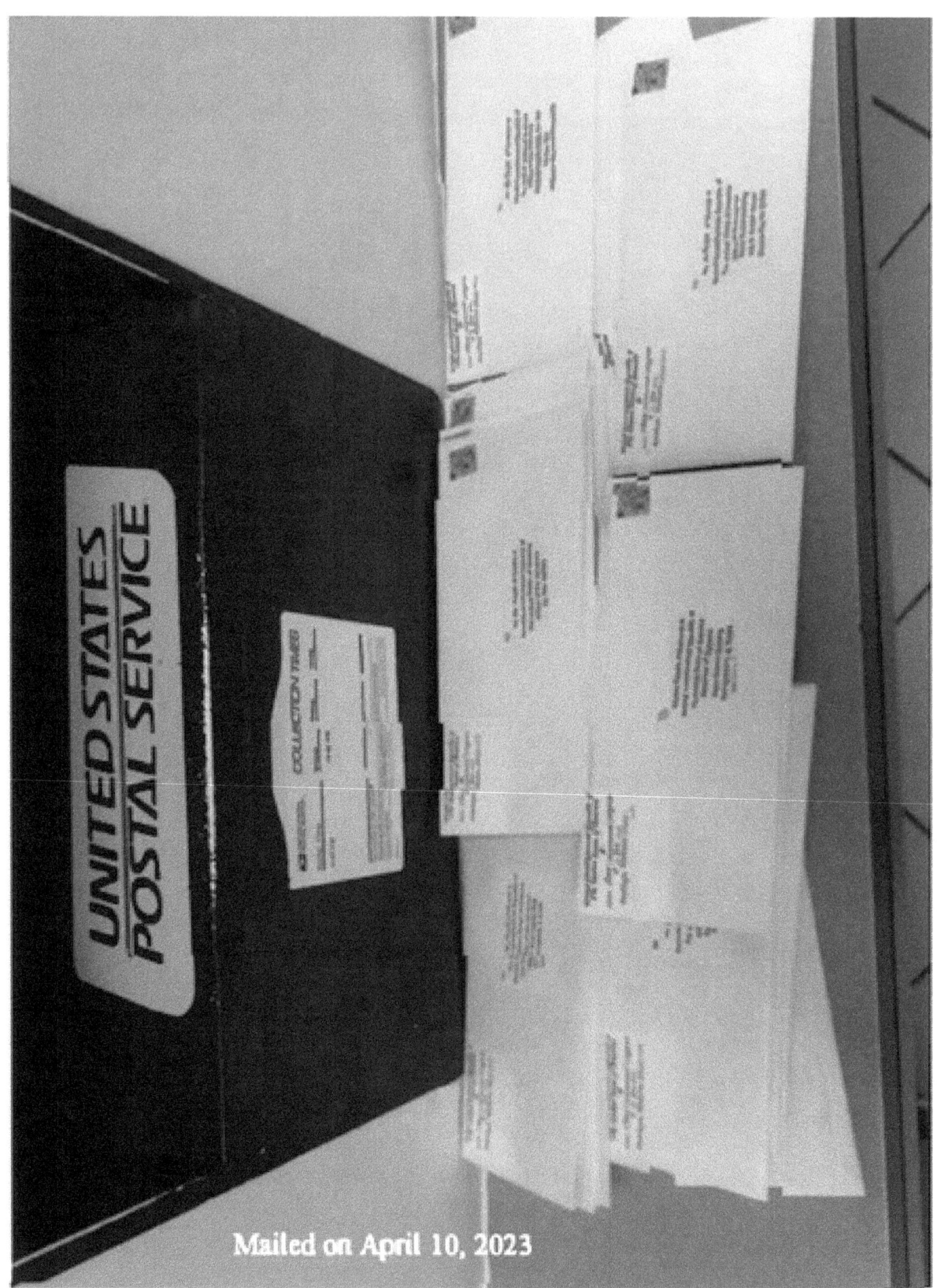

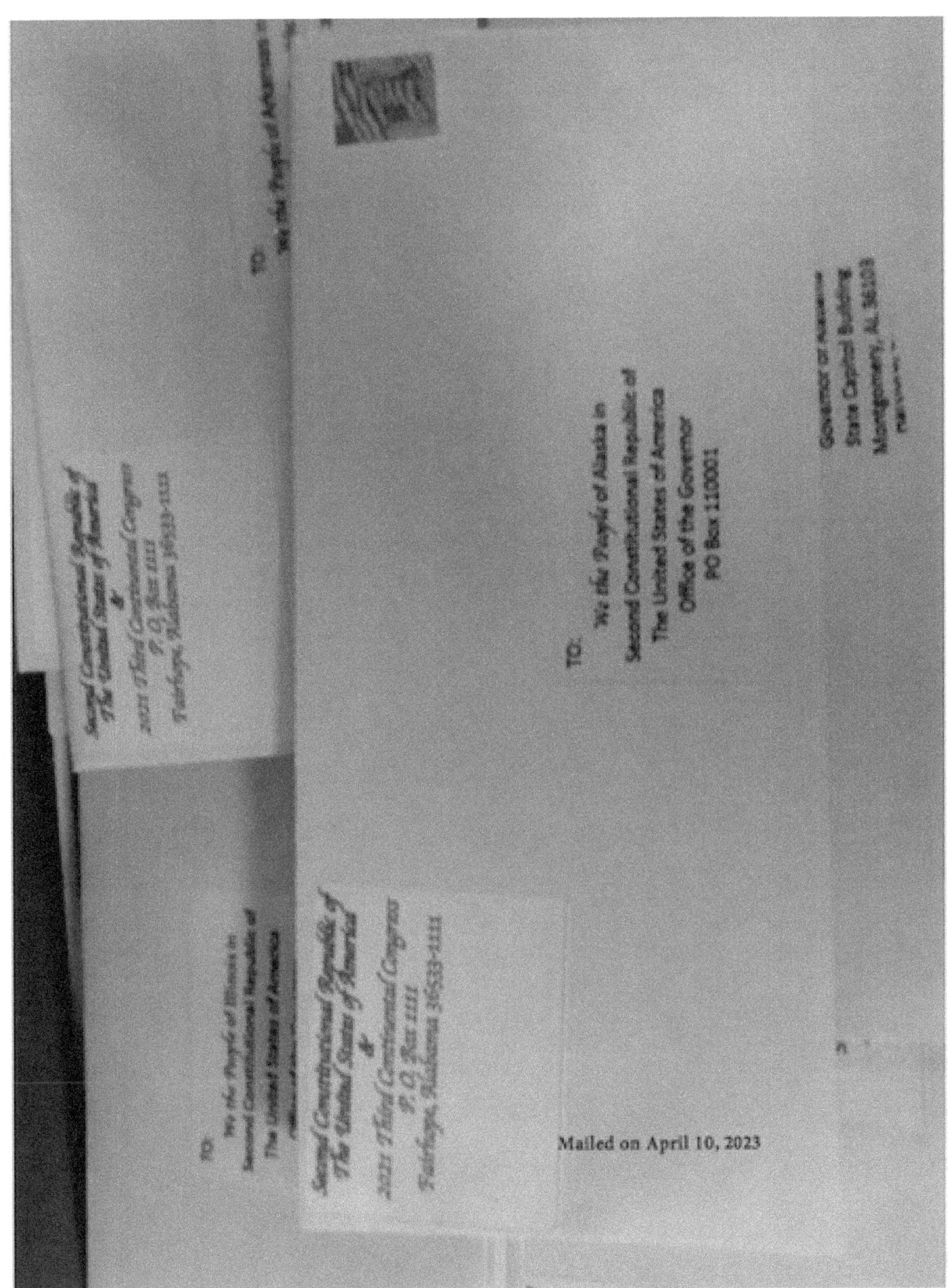

Mailed on April 10, 2023

Greetings Honorable University President:

The Lord has revealed to me, when He will fulfill his promise to return on His third day or the start of His thousand years reign of peace at your gate (2 Peter 3:8-9, 2 Peter 3:1-18, Genesis 6:3 and Daniel 2:35). Same as the Jewish descendance saved by God's closing of the Red Sea, all God's creations are being saved by a "Crimson (Red) Tide," metaphorically speaking, like in the Bay of Fundy Minas Basin, where the height of the tide can reach 16 meters (53ft). When the Red Sea suddenly closed, both good and evil people died. This time, God refers to a "Crimson Tide" to illustrate He is giving everyone time to repent, not wishing that anyone should perish. It was Satan's deception that the Bible does not say when our Lord Jesus Christ's shall return. By the Grace of God and through God's anointed modern day profits, God told the world the exact date of Christ's return. Since the last prayer, at the end of the 1st Passover and end of Genesis 6:3's 50th Jubilee, God's beloved Jewish children have ended their last prayer with: Next year in Jerusalem. This is that year, on June 3, 2023, the beginning of Genesis 6:3's 121st Jubilee.

Revelation 20:2-7 ESV

The Thousand Years

20 Then I saw an angel coming down from heaven, holding in his hand the key to the bottomless pit[a] and a great chain. 2 And he seized the dragon, that ancient serpent, who is the devil and Satan, and bound him for a thousand years, 3 and threw him into the pit, and shut it and sealed it over him, so that he might not deceive the nations any longer, until the thousand years were ended. After that he must be released for a little while.

The "Crimson Tide" came in the form of the digital computers, blockchain Computers and the StarLink satellite communication system. Satan's dominions owned the patents and copyrights and were going to destroy Humanity with them. But by the Grace of God, God has made them usable now and they will be obsolete when the Christ Consciousness Base 2- Base 3 Spherical - Q:ube Quantum Computers are built.

The enclosed book, "The Story of Our Life, Based on A True Life.", contains copyrighted technology to design, build and program Christ's Quantum Computer.

All existing digital devices will continue to be used and will temporally interact with Christ's Quantum Computers. Even when all existing digital devices are being used, they will not or ever will be any artificial intelligence (AI) software be used, no personal information collected or used, device or system hacking, sending or receiving unsubscribed information and no money fraud, illegal transactions or theft of US Treasury gold backed currency credits.

The Second Constitutional Republic of The United States of America (USA) will be contracting the design, construction and programing of hand held and mainframe Christ Quantum Computers and Satellite communications systems to be securely used by all nations on earth. The designers and subsequent builders shall keep all royalty rights. However, the USA shall keep all patents and copyrights to maintain the security and trust of everyone on earth.

The true Quantum technology in my book and video, known by Dr. Albert Einstein and Nikola Tesla, shall be taught to everyone on earth. To get things started, I am offering the Universities listed below the first opportunity to spread the knowledge in their native language. For example, if the Hebrew University of Jerusalem does not sign the attached contract by June 3, 2023, I will offer the opportunity to another Hebrew University. After June 3, 2023, I anticipate getting request to participate from every country in the world.

Free ebook and video downloads are available at www.piisthree.com and www.piis3.com under "Our Products."

Please mail signed contracts to James N. Akins, Jr., PO Box 91, Fairhope, Alabama 36533-0091.

God Bless everyone,

James N. Akins, Jr.

CC:

1. University President
   Department of Physics
   National and Kapodistrlan University of Athens University Campus
   GR-157 84 Z0grafou, Athens
2. University of Vienna, President
   Universitätsring 1
   1010 nna
3. The Alexander Silberman Institute of Life Sciences
   The Hebrew University of Jerusalem
   Edmond J. Safra Campus - Givat Ram
   Jerusalem 9190401
   Israel
4. The University of Alabama, President
   Tuscaloosa, AL 35487
5. The University of Science and Technology Beijing, President
   30 Xueyuan Road
   Haidian District, Beijing 100083, China
6. University of Tokyo, President
   3-8-1 Komaba Meguro-ku
   Tokyo 153-8914, Japan
7. University of Mumbai, President
   Mahatma Gandhi Road
   Fort, Mumbai, Maharashtra 400032.
8. Science & Technology
   National Polytechnic Institute, President
   Av. 510 1000
   Pueblo de San Juan de Aragón, Gustavo A. Madero
   Mexico City, 07480
9. Auburn University
   Auburn, Alabama
10. University of New Mexico
    2500 Campus Blvd NE
    Albuquerque, NM 87131
11. Cairo University
    1 Gamaa Street, Giza, Egypt
    Postal Code: 12613

1. This contract is to grant to 0ne (1) university in each language in the world, the publishing rights to grant subsequent independent published BOOKS in the "The Story of Our Life, Based on A True Life" book series. After the initial contract signing between THE UNIVERSITY and JAMES N. AKINS, JR. an amended signature page shall be add each time an independent BOOK AUTHOR is added to the BOOK SERIES. All pervious BOOK SERIES AUTHORS must approve and sign acceptance of the NEW BOOK AUTHOR.

2. Each BOOK AUTHOR shall be an independent publisher, same as π = 3 BOOKS, and NO BOOK AUTHOR and JAMES N. AKINS, JR. shall have ANY LEGAL COPYRIGHT CLAIMS on another's intellectual property or book sells royalties.

3. To be accepted as a BOOK SERIES AUTHOR:

    1. The first prospective BOOK 1 AUTHOR shall submit an audio book published under my publishing company, π = 3 BOOKS including my International Standard Book Number (ISBN) for Audio Book. The reading of "The Story of Our Life, Based on A True Life", including the Copyright page, shall be in the BOOK 1 AUTHOR'S native language.

        A. The Audio book recording shall be in the prospective BOOK 1 AUTHOR's native language. The Narrator shall state what page number they are going to read. Most pages can be simply read. However, the following listed pages require special instructions.

            1. Narrator says: Page 1. The Story of Our Life, Based on A True Life. The Message in a Bottle - Norma, I now Know. Picture of Typed Letter out of bottle and the handwritten Message from a bottle.

            2. Narrator says: Page 7. Newspaper Article.

            3. Narrator says: Page 15 and 16. God's Law of Relativity Diagramed of 2π divided by 6. E=Mc squared.

            4. Narrator says: Page 17, 18, 19 and 20. Fabric of Time and Space. God's Law of Relativity Diagramed of 2π divided by 6. E=Mc squared.

            5. Narrator says: Page 22. Human Brain Q:bit Quantum Computer. God's Law of Relativity Diagramed of 2π divided by 6. E=Mc squared.

            6. Narrator says: Page 24. Leonard DiVinci, Vitruvian Man in E=Mc squared.

            7. Narrator says: Page 25. Leonard DiVinci, Vitruvian woman in E=Mc squared.

            8. Narrator says: Page 32. Leonard DiVinci, Vitruvian Man in E=Mc squared and 2π divided by 5 and 2π divided by 6.

            9. Narrator says: Page 33. Leonard DiVinci, Vitruvian woman in E=Mc squared and 2π divided by 5 and 2π divided by 6.

10. Narrator says: Page 50. Snow Flakes formed on Christ's frequency 2π divided by 6.

11. Narrator says: Page 51. Unknown artist- Sacred Geometry art examples of E=Mc squared and 2π divided by 5 and 2π divided by 6.

4. All BOOK SERIES AUTHOR shall post my Copyright page after the BOOK AUTHOR's Copyright page, ATTACHED EXHIBIT I.

5. All BOOK SERIES AUTHOR shall be provided the BOOK SERIES cover page, ATTACHED EXHIBIT II.

6. All BOOK SERIES AUTHOR shall use ATTACHED EXHIBIT II and print the BOOK NUMBER, BOOK TITLE and AUTHOR's NAME in BOOK AUTHOR's Native Language, shown in yellow letters, EXHIBIT III.

Make two (2) original copies. Sign both as University President, date stamp and seal with University Seal. Mail to James N. Akins, Jr., PO Box 91, Fairhope, Alabama 36533-0091.

EXHIBIT I: Copyright Page from "The Story of Our Life, Based on A True Life. By Bubba "His X. Mark" Twain.

EXHIBIT II: BOOK SERIES COVER PAGE, EXAMPLE.

EXHIBIT III: BOOK SERIES COVER PAGE with Native Language BOOK NUMBER, BOOK TITLE, and BOOK AUTHOR, example in yellow.

James N. Akins, Jr.

University President

The Story of our Life, Based on a True Life.
All Rights Reserved.
Copyright © 2022 Bubba "His X. Mark" Twain
v2.0 r2

Previous Printings 2019, 2020, 2021, 2022

The opinions expressed in this manuscript are solely the opinions of the author and do not represent the opinions or thoughts of the publisher. The author has represented and warranted full ownership and/or legal right to publish all the materials in this book.

This book may not be reproduced, transmitted, or stored in whole or in part by any means, including graphic, electronic, or mechanical without the express written consent of the publisher except in the case of brief quotations embodied in critical articles and reviews.

Paperback ISBN: 978-0-578-26233-8
Hardback ISBN: 978-0-578-26237-6

π – 3 Books

Library of Congress Control Number: 2022904632

New and revised text © 2022 by James N. Akins, Jr. Certificate of Registration Number TX 9-092-173 and TX 9-121-305 Cover Illustration by Victor Guiza © 2022 by James N. Akins, Jr.

Artwork and text - © 2018 by James N. Akins, Jr.
Certificate of Registration Number TXu 2-130-773

2 D Artwork, MC2 Defined - © 2018 by James N. Akins, Jr.
Certificate of Registration Number VAu 1-398-340

PRINTED IN THE UNITED STATES OF AMERICA

2021 Third Continental Congress
Office of the Speaker of the House of Representative
Carpenters' Hall
Philadelphia, Pennsylvania
October 1, 2023

U. S. Department of The Treasury, Director
National Payment Integrity and Resolution Center
P. O. Box 51315
Philadelphia, Pennsylvania 19115-6314

Greetings Director,

We the People of the Second Constitutional republic of The United States of America and in accordance to the 1776 Declaration of Independence and 1787 United States of America Constitution financial and economic retribution for the money Extorted from all We The People (a.k.a "American Tax Payers") since 1776 by the following Unconstitutional and illegal organizations and individuals (Defendants):

Violates Article I, Section 8, Clause 1 and/or Article XI of the 1787 United States of America Constitution.

1. Internal Revenue Service (IRS),
2. Bankrupted United States of America, INC. (USAINC),
3. Bankrupted United Nations, INC. (UNINC),
4. UNINC US Office of Personnel Management (OPM) (i.e. FBI, CIA, Congress, IRS, President, Supreme Court Justices, etc.)
5. British Accreditation Registry-Crown Temple British Maritain Law (B.A.R.)
6. The Central Bank of the USAINC is the Private Federal Reserve System, created by Congress in 1913.
7. Private Western Central Vatican Bank and Central Bank of London.
8. North Atlantic Treaty Organization (NATO)

Attached is EXHIBIT V, to be added to my claim.

Praise God for His gifts of Life, Liberty, and the Pursuit of Happiness to all mankind.

James N. Akins, Jr. © Without Prejudice

EXHIBIT VI

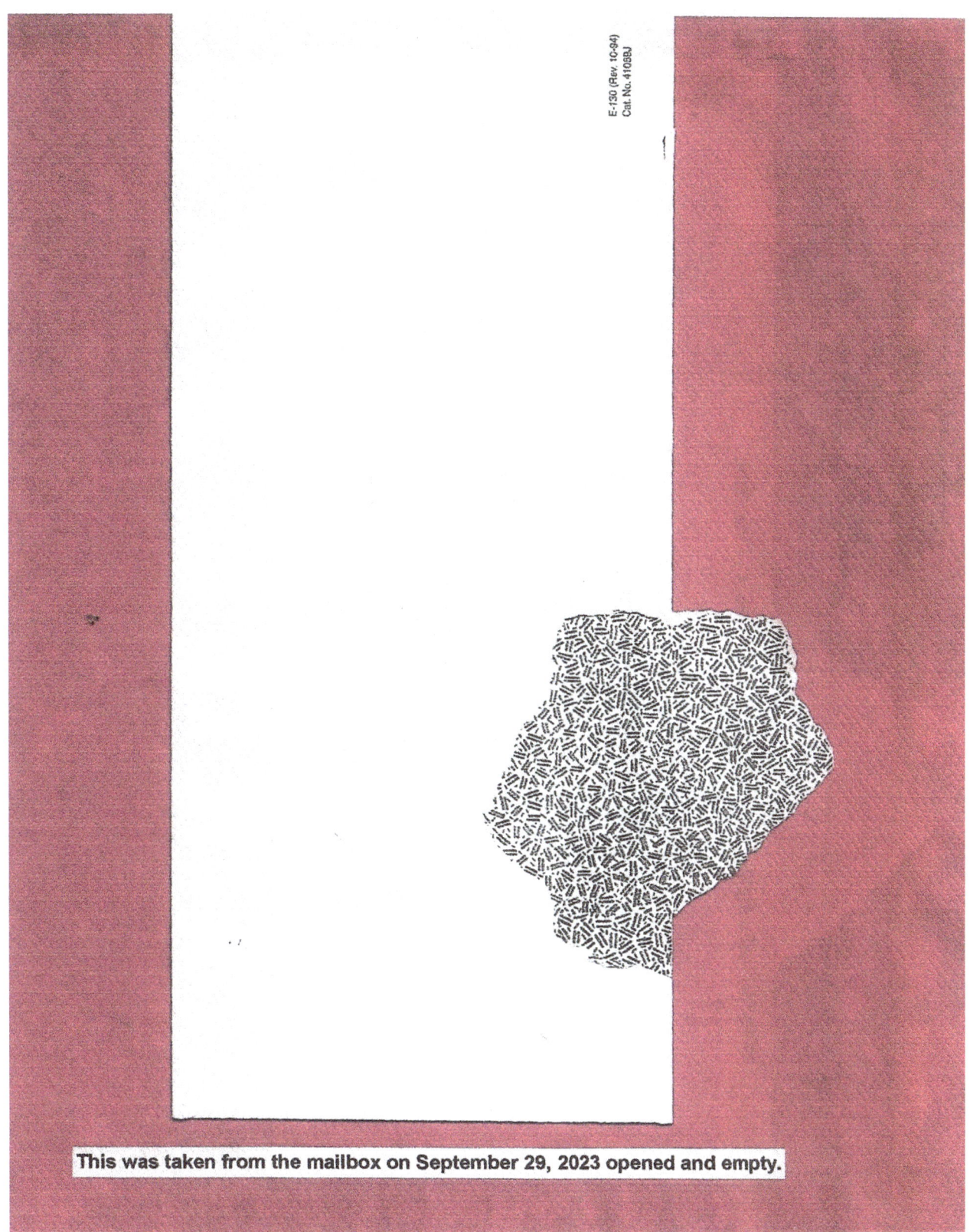

This was taken from the mailbox on September 29, 2023 opened and empty.

By the written word of God and the signs in the firmament of Heaven, it shall be known that no one, but everyone shall know the day and hour Lord Jesus Christ returns to start His 1,000 years reign of peace at our gates. Unlike the surprise closing of the Red Sea, God's justice shall be served, like the known time of a coming Crimson Tide, that shall destroy all that did not repent and the casting of satan into a pit for 1,000 years.

The end of the 120th Jubilee, described in Genesis 6:3 and Daniel 2:3, is marked by the end of the winter solstice, precicely between December 21, 2022 at 24:00 hour and December 22, 2022 at 00:00 hour Jerusalem time or the same day and hour Jesus was conceived 2,022 years ago. Jesus' birthday was at the end of the fall equinox, precicely between September 21, 2022 at 24:00 hour and September 22, 2022 at 00:00 hour Jerusalem time.

As God promised, the expiration date of the Babylonian Kingdom and the Great Restoration Jubilee shall be fulfilled in one day, marked by the end of the winter solstice, precicely between December 21, 2023 at 24:00 hour and December 22, 2023 at 00:00 hour Jerusalem time.

BOOK I CHAPTER IIII – QUBE BYTES:
THE STORY OF OUR LIFE BASED ON A TRUE LIFE THE $\pi = 3$ BOOKS SERIES

BOOK TITLE

AUTHOR'S NAME

The STORY of OUR LIFE, BASED on A TRUE LIFE.
The π = 3 BOOKS SERIES
BOOK 1
CHAPTER 1111 - QUBE BYTES: * FOR THE RECORD!
* THE LIBRARY of CONGRESS CONTROL NUMBER: 2022904632
All Rights Reserved.
V3.1 Copyright © 2023 James N. Akins, Jr.

The opinions expressed in this manuscript are solely the opinions of the author and do not represent the opinions or thoughts of the publisher. The author has represented and warranted full ownership and/or legal right to publish all the materials in this book.

This book may not be reproduced, transmitted, or stored in whole or in part by any means, including graphic, electronic, or mechanical without the express written consent of the publisher except in the case of brief quotations embodied in critical articles and reviews.

Paperback ISBN: 979-8-218-96222-7
Hardback ISBN: 979-8-218-96232-6

Cover Art, Illustrations, Unique font selection, Words/phrase meaning of:

"The Story of Our Life, Based on A True Story.
The π = Books Series
** BOOK # - *** Volume #
CHAPTER IIII - QUBE BYES:" /Copyright ©James N. Akins, Jr. 2023.
TITLE
AUTHOR

* FOR REFERENCE ONLY
The actual Library of Congress Control Number: 2022904632 (LOCCN: 2022904632) File is in the Library of Congress and contains all editions of the book: "The Story of Our Life, Based on A True Story." (As well as, original documents featured in the book.). The use of LOCCN: 2022904632 in the title of this unique special International Standard Book Number (ISBN:) book is for reference only. In all future "The π = 3 BOOKS Series" books, the actual "LOCCN: 2022904632 File" in the Library of Congress, represents the physical embodiment of the "lead character" in the series, same as the "Rings" in J.R.R. Tolkiens' "Lord of the Rings" series.
** BOOK #
All English Language books, in the series, shall be BOOK 1, with 1 TITLE, 1 AUTHOR, and unique special International Standard Book Number (ISBN:) per book TITLE.

All Foriegn Language books, in the series, shall be a different BOOK # starting with "BOOK 2- Volume I", with 1 TITLE, 1 AUTHOR, and unique special International Standard Book Number (ISBN:) per book TITLE (i.e. Hebrew Language is BOOK 2 - Volume I, Japaneese Language is BOOK 3 - Volume I, and so on.) Each unique foreign language BOOK # and VOLUME TITLE, AUTHOR and CONTENT shall be written in only the native language of the unique BOOK # and AUTHOR.

*** VOLUME #
After each unique "BOOK # - VOLUME I, TITLE, AUTHOR," 100 % of all previous authors, in a unique "BOOK #" series, shall approve all subsequent unique "VOLUME #" AUTHOR. This is a new, domestic and foreign, book series that is only authorized reproduction and sell shall be in writing from James N. Akins, Jr..

"It has been a long standing understanding the spelling of cube and cubit is correct. In God's Quantum geometry technology, a Qube is equal sides of 1= 9 Qubes (9/9) or 10 Qubebits (!0/10). In God's Quantum Computer technology, there are Qube Bytes. Unlike the present math, the largest whole number is 1, not an infinite number of whole numbers as we are taught now. All this is made clear in my book." /Copyright ©James N. Akins, Jr. 2023

**** © Newton, Isaac, Philosophiae Naturalis Principia Mathematica("Mathematical Principles of Natural Philosophy"), London, 1687; Cambridge, 1713; London, 1726. (Pirated versions of the 1713 edition were also published in Amsterdam in 1714 and 1723*
π = 3 Books
Library of Congress Control Number: 2023915883

PRINTED IN THE UNITED STATES OF AMERICA

Internal Revenue Service
Stop
Cincinnati OH 45999

Official Business
Penalty for Private Use, $300

RETURN TO SENDER
I, JAMES N. AKINS, JR., DO HEREBY
REFUSE TO CONTRACT WITH THE DEFUNKED
PRIVATE CORPORATION IRS, INC. THIS MAIL IS
IN VIOLATION OF SECOND CONSTITUTIONAL
REPUBLIC OF THE UNITED STATES OF
AMERICA US POSTAL SERVICE. LAW

James Akins
PO Box 1111
Finchpen, AL 36533-1111

EXHIBIT V

ORIGINAL AT:

U. S. Department of the Treasury
National Payment Integrity and
Resolution Center
P. O. Box 51315
Philadelphia, Pennsylvania
19115-6314

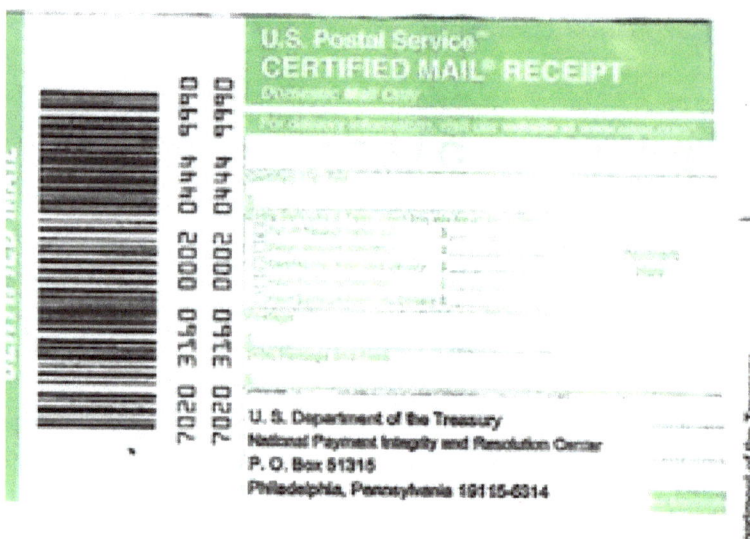

Final Testament – Rise of the Holy Spirit.

Biblical milestone dates:

December 21, 2022 – 24:00:00 midnight (Jerusalem Time)

- Blowing of the 4th Trumpet, marking 15 minutes until the end of "The Magador Bull Fight". In The Magador, as told by Juan O'Savin , 15 Minutes in Heaven was December 21, 2022 – 00:00:00+ plus 91.5 days of 366 days or Spring Solstice. Blowing of the 5th Trumpet on March 20th, 2023 – 12:00:00. Then in Revelation 8:1 Jesus breaks the seventh seal at the blowing of the 6th Trumpet on April 7th, 2023 at 18.00.03, 2023 years after His Crucifixion. John says that, after this seal was broken, "there was silence in heaven for about half an hour." This silence immediately follows the loud and jubilant songs of heavenly worship in chapter 7, making the absolute, sudden silence even more dramatic. There is something about the seventh seal that stops every mouth and silences all of heaven. The silence is broken by the Beginning of the Reign of our Lord Jesus Christ.

- End of the 120th Jubilee (6,022 years since God's Creation of Adam) (Genesis 6:3)

- God destroys the 4th Evil Kingdom, … it shall break in pieces and consume all these kingdoms, and it shall stand for ever (Daniel 2:35, Daniel 2:40 – 2:44). The last Kingdom is an everlasting Kingdom (Daniel 7:18; Daniel 7:27).

December 22, 2022 – 00:00:00+ past midnight (Jerusalem Time)

- 2022th Birthday of Lord Jesus Christ

- Beginning of Lord Jesus Christ's Reign on His 3rd day (Daniel 2:35), … "But do not overlook this one fact, beloved, that with the Lord one day is as a thousand years, and a thousand years as one day. The Lord is not slow to fulfill his promise as some count slowness, but is patient toward you, not wishing that any should perish, but that all should reach repentance." (2 Peter 3:8-9)

Revelation 8:1- Revelation 20:2-7 ESV

The Thousand Years Reign of *Our* Lord Jesus Christ

20 Then I saw an angel coming down from heaven, holding in his hand the key to the bottomless pit[a] and a great chain. 2 And he seized the dragon, that ancient serpent, who is the devil and Satan, and bound him for a thousand years, 3 and threw him into the pit, and shut it and sealed it over him, so that he might not deceive the nations any longer, until the thousand years were ended. After that he must be released for a little while.

April 7, 2023 – Sunset 6PM+ 3 seconds (Jerusalem Time)

- As the apostle John relates an extended vision of God's throne room, he describes the Lamb on the throne who is handed a scroll with seven seals (Revelation 6). The Lamb (Lord Jesus Christ) proceeds to break open the seals one by one. After each seal is opened, a judgment occurs on earth. Then, "when he opened the seventh seal, there was silence in heaven for about half an hour" (Revelation 8:1) (Silence in heaven for about half an hour, because only our Lord Jesus Christ knows the exact time of his second coming and the sounding of the 7th Trumpet. About 30 Minutes in Heaven - April 7, 2023 – Sunset 6PM+ 3 seconds (Jerusalem Time) plus about 183 days or around October 09, 2023. After the half hour of silence, the seven trumpet judgments begin (Revelation8:6—9:21; 11:15–19).

- There are 120 Jubilees in Genesis 6:3, each 50 years ends with a 1 year Jubilee (Group of 7 years, like a week, times 7 equals 49 years plus 1 year of Jubilee equals 50 years. 6,022 years since God's Creation of Adam. The 22 years accounts for the spiral of 6000 years plus spiraling of time between 366 days to 365 days in 4 years, then 365 days to 366 days in 4 years.) (Genesis 6:3)

- Our Lord Jesus Christ was born at the start of the 80th Jubilee December 22, 0000 just after midnight 00:00:00+. The end of the 120th Jubilee (6,022 years since God's Creation of Adam) (Genesis 6:3) was on December 21, 2022 at 24:00:00

- This important feast gets its name from the fact that it starts seven full weeks, or exactly 50 days, after the Feast of Firstfruits. Since it takes place exactly 50 days after the previous feast, this feast is also known as "Pentecost" (Acts 2:1), which means "fifty." At the end of time end times, the Feast of the Pentecost takes on the marking of the end of negative 120 Jubilees (7 X 7 = 49 + 1 = 50 years x 120 Jubilees = 6000 or 1000 years to Christ is like 1 day. The same as God created the earth in 6 days and rested on the 7th day. After the first Passover, the first day of the Feast of the Pentecost is the day after the 7th day of the Passover.

2022 (-)120th Jubilee – 1st day of 2022 Passover and Good Friday was on April 15, 2022 – Sunset -18:00:03.

2022 (-)120th Jubilee – 7th day of 2022 Passover was on April 22, 2022 – Sunset.

2022 (-)120th Jubilee – 1st day of 2022 Feast of the Pentecost started on April 22, 2022 – Sunset.

2022 (-)120th Jubilee – 50th day of 2022 Passover or Day of Jubilee was on June 5, 2022 – Sunset.

- Our Lord Jesus Christ was born at the start of the 80th Jubilee December 22, 0000 just after midnight 00:00:00+. The end of the 120th Jubilee (6,022 years since God's Creation of Adam) (Genesis 6:3) was on December 21, 2022 at 24:00:00 at midnight.

- The start of the (+)1st Jubilee was on December 21, 2022 at 00:00:00+ after midnight (Daniel 2:35, Daniel 2:40 – 2:44). The last Kingdom is an everlasting Kingdom (Daniel 7:18; Daniel 7:27).

2023 (+)1st Jubilee – 1st day of 2023 Passover was on April 5, 2023 – Sunset.

2023 (+)1st Jubilee – 1st day of 2023 Good Friday was on April 7, 2023 – Sunset - 18:00:03.

2023 (+) 1st Jubilee – 7th day of 2023 Passover was on April 12, 2023 – Sunset.

2023 (+)1st Jubilee – 1st day of 2023 Feast of the Pentecost started on April 12, 2022 – Sunset.

2023 (+)1st Jubilee – 50th day of 2023 Passover or Day of Jubilee was on June 2, 2023 – Sunset.

Please go to www.piisthree.com or www.piis3.com to download FREE eBook "The Story of Our Life, Based on A True Life" and watch the videos.

Through the name of our Lord Jesus Christ: God's will is done!

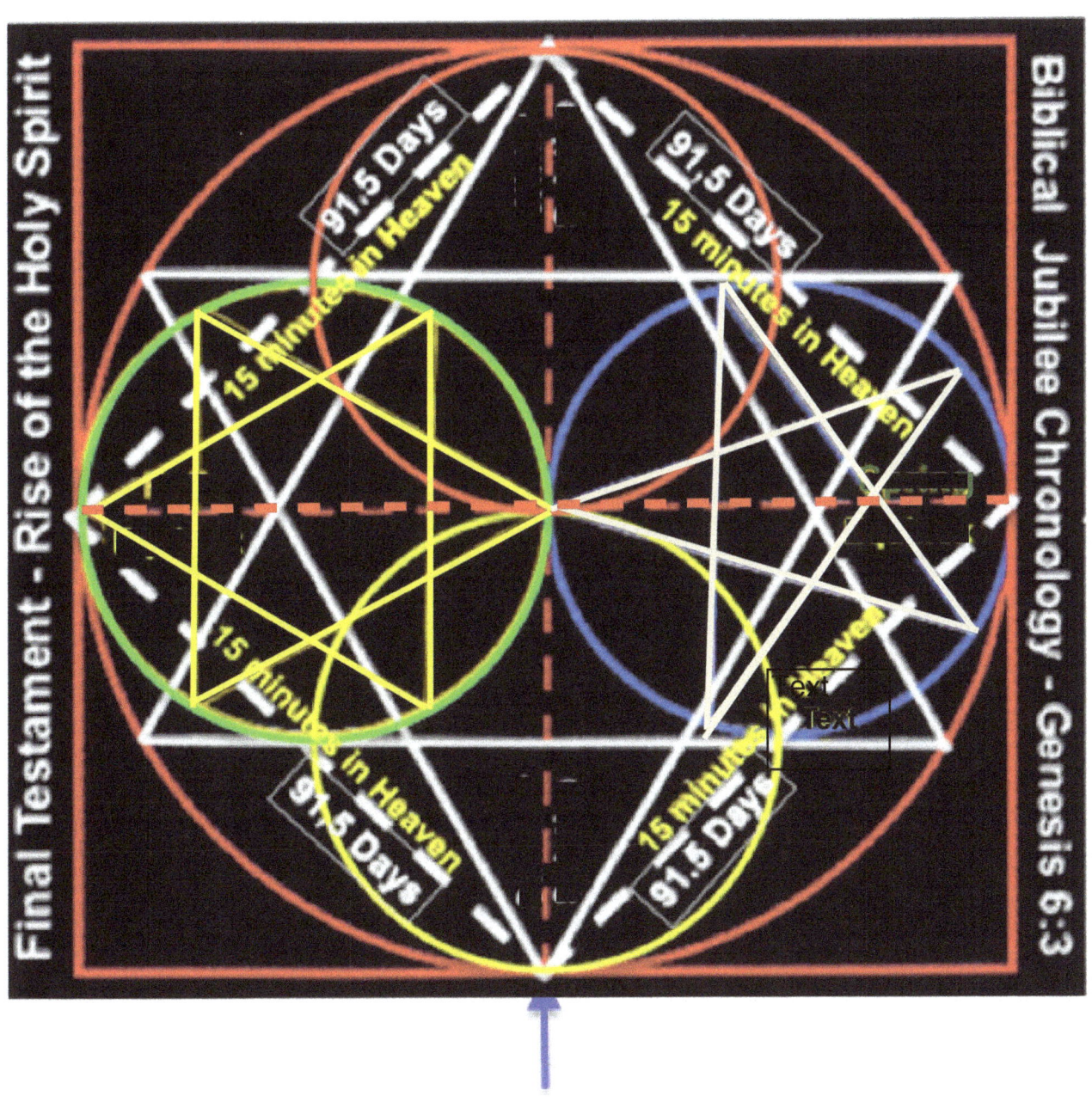

Final Testament – Rise of the Holy Spirit.

Our Lord Jesus Christ was born at the start of the 80th Jubilee December 22, 0000 just after midnight 00:00:00+. The end of the 120th Jubilee (6,022 years since God's Creation of Adam) (Genesis 6:3) was on December 21, 2022 at 24:00:00. The 2022th birthday of Our Lord Jesus Christ, on December 22, 2022 at 00:00.00+, marks the start of the 1st Positive (+) Jubilee of the Final Testament – Rise of the Holy Spirit.

I must apologize to Sir Isaac Newton for the uneducated statements I made in my book about him and his life's work. Sir Newton and I had three things in common; 1. We are just men of science that knew God is real and the Holy Bible is the written words of God. When the Holy Bible is decoded, it reveals the true Quantum mathematatica, astrology, Quantum Computers and software and physics of God's creations. Sir Isacc Newton sums up what I am saying in his two books: ****"Philosophiae Naturalis Principia Mathematica." In short, my perception of Newton's Calculus, as a lie or evil was wrong. Because, God wanted Satan and his dominions to think Newton's Calculus, as I point out in my book, was their means to rule over and destroy God's Creations through their creations; digital computers and Blockchain Quantum computers. They were wrong: God's Holy Bible WON!; 2. Both Sir Isaac Newton and I both came to the same conclusion, the Holy Bible recorded the True Advanced Technology that existed during the days of Moses and Noah. Sir Isaac Newton, Dr. Albert Einstein and Nikola Tesla were geniuses. However, most importantly they were profits in their days that 100% followed God's will through their Faith. Faith in God's will and Our Lord Jesus Christ is what we four guys have in common. and; 3. Our mothers do not understand us and God's plan for our life's contributions to Humanity. My mother thinks I am crazy.

**** © Newton, Isaac, Philosophiae Naturalis Principia Mathematica("Mathematical Principles of Natural Philosophy"), London, 1687; Cambridge, 1713; London, 1726. (Pirated versions of the 1713 edition were also published in Amsterdam in 1714 and 1723.)

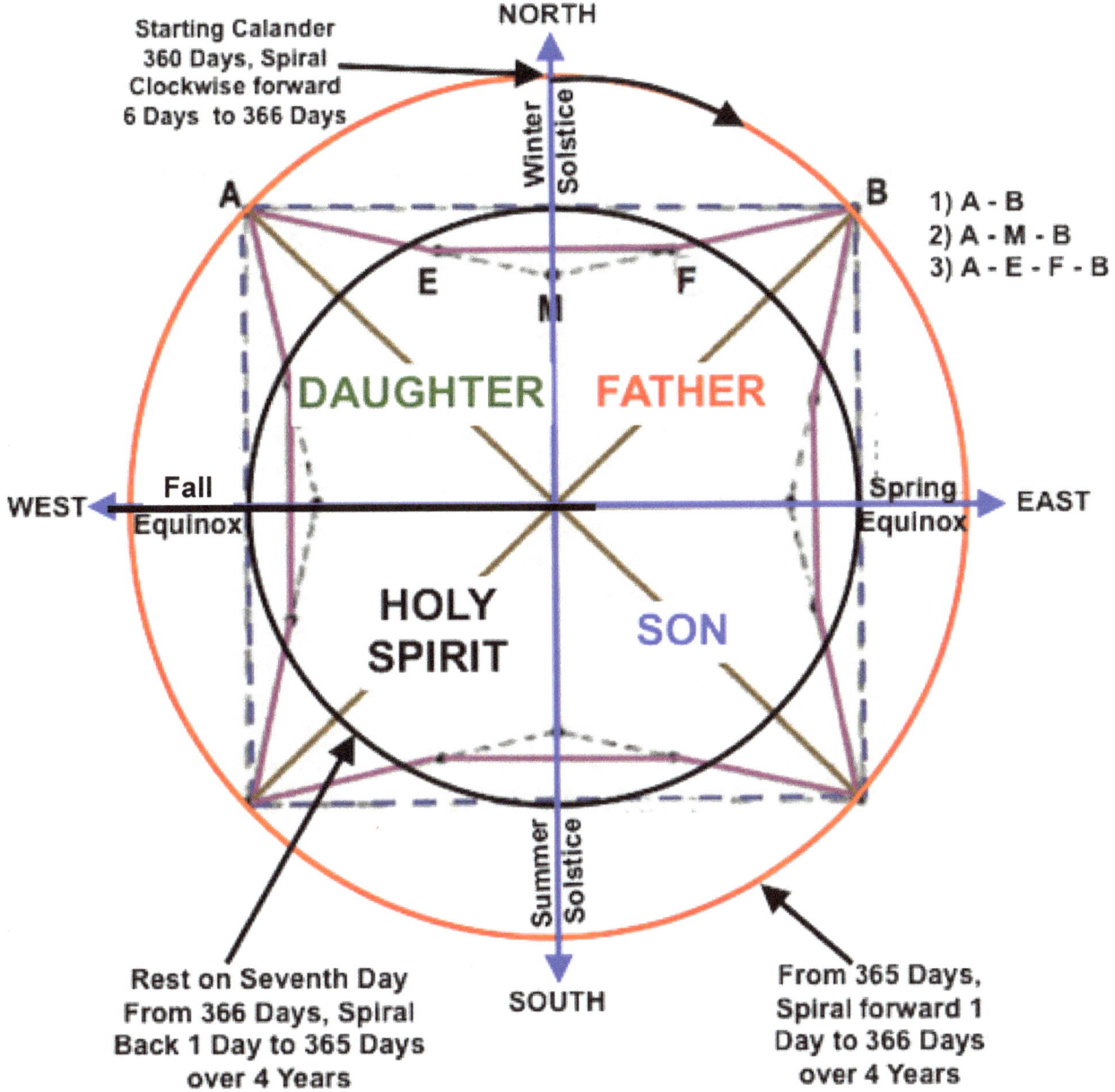

Pyramid of Khufu
Top Down View of Great Pyramid, Egypt

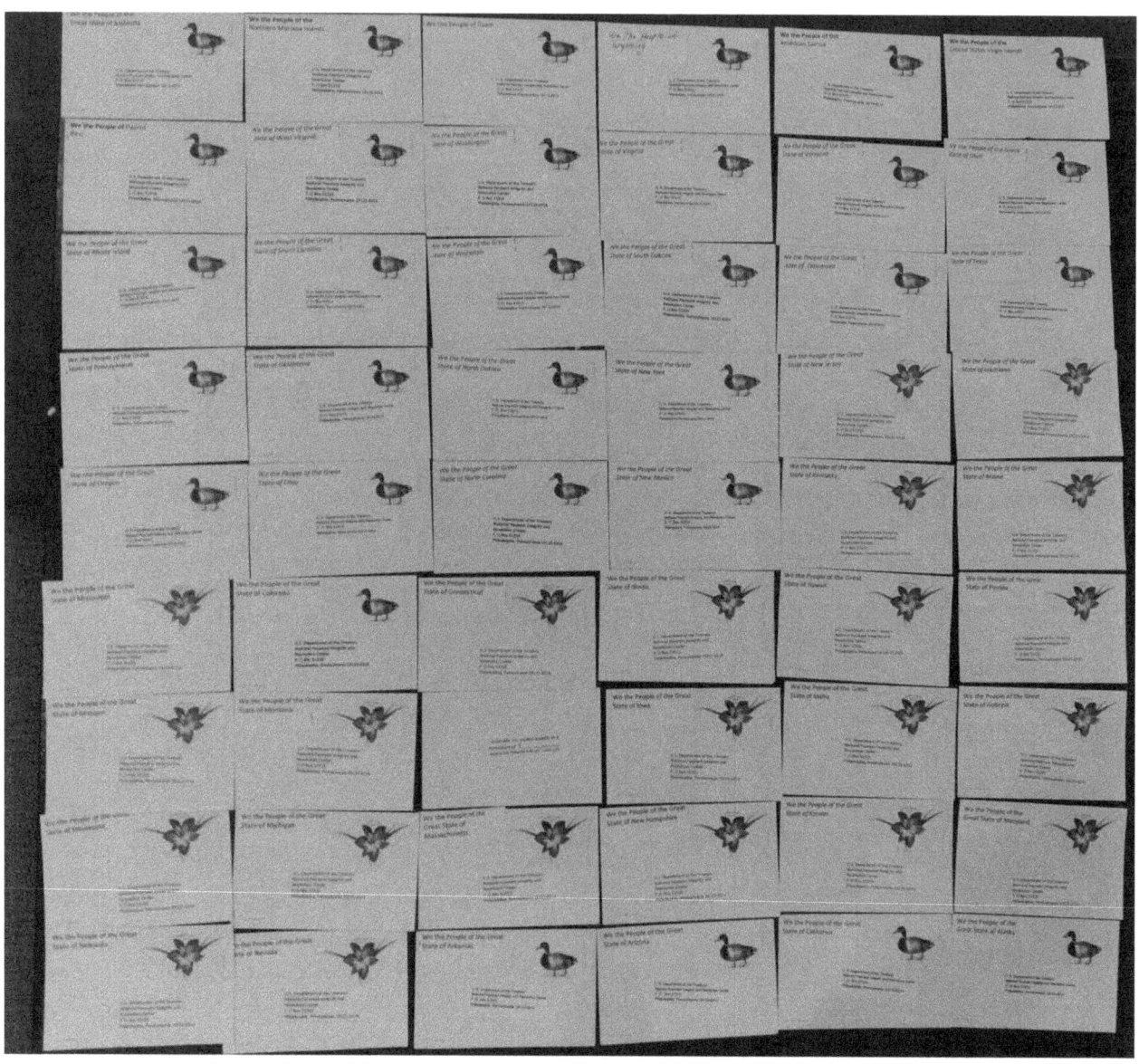

August 15, 2023, 50 states and 5 American Territories post cards mailed to: U. S. Department of the Treasury, National Payment Integrity and Resolution Center, P. O. Box 51315, Philadelphia, Pennsylvania 19115-6314 with each initial claim of $15 Trillion in resolution for a total of $825 Trillion. All United States of America Citizens are encouraged to submit their family's resolution claim back to 1812 times 10.

2021 Third Continental Congress
Office of the Speaker of the House of Representative
Carpenters' Hall
Philadelphia, Pennsylvania
August 28, 2023

U. S. Department of The Treasury, Director
National Payment Integrity and Resolution Center
P. O. Box 51315
Philadelphia, Pennsylvania 19115-6314

Greetings,

I am James N. Akins Jr., Speaker of the House of Representative of The 2021 Third Continental Congress (TCC). As you are aware, We the People of the Second Constitutional Republic of The United States of America and in accordance to the 1776 Declaration of Independence and 1787 United States of America Constitution claiming financial and economic retribution for the money Extorted from all We The People (a.k.a "American Tax Payers") since 1915, by the following Unconstitutional and illegal organizations and individuals (Defendants):

Violates Article I, Section 8, Clause 1 and/or **Article XI of the 1787** United States of America Constitution.

1. Internal Revenue Service (IRS),
2. Bankrupted United States of America, INC. (USAINC),
3. Bankrupted United Nations, INC. (UNINC),
4. UNINC US Office of Personnel Management (OPM) (i.e. FBI, CIA, Congress, IRS, President, Supreme Court Justices, etc.)
5. British Accreditation Registry-Crown Temple British Maritain Law (B.A.R.)
6. The Central Bank of the USAINC is the Private Federal Reserve System, created by Congress in 1913.
7. Private Western Central Vatican Bank and Central Bank of London.
8. North Atlantic Treaty Organization (NATO)

The Second Constitutional Republic of The United States of America, Attorney General shall coordinate with the Governor of each of the 50 United States during the resolution of this Extortion Case and provide a progress report weekly.

By September 15, 2023, you are hereby directed to report the status of claims to date to Congress;

2021 Third Continental Congress
Speaker of the House of Representatives
P. O. Box 1111
Fairhope, AL 36533-1111
Mr.akins21@yahoo.com

Praise God for His gifts of Life, Liberty, and the Pursuit of Happiness to all mankind.

James N. Akins, Jr. © Without Prejudice
2021 Third Continental Congress
Speaker of the House of Representatives

Chapter 2

The Story of Our Life, Based on a True Life., by Bubba "His X. Mark" Twain

הספר האחרון של הברית החדשה, רוב שוב אדי ושינו מהשיח

THE STORY OF OUR LIFE, BASED ON A TRUE LIFE

BUBBA "HIS X. MARK" TWAIN

**The STORY of OUR LIFE,
based on a TRUE LIFE.**

The Story of our Life, Based on a True Life.
All Rights Reserved.
Copyright © 2022 Bubba "His X. Mark" Twain
v2.0 r2

Previous Printings 2019, 2020, 2021, 2022

The opinions expressed in this manuscript are solely the opinions of the author and do not represent the opinions or thoughts of the publisher. The author has represented and warranted full ownership and/or legal right to publish all the materials in this book.

This book may not be reproduced, transmitted, or stored in whole or in part by any means, including graphic, electronic, or mechanical without the express written consent of the publisher except in the case of brief quotations embodied in critical articles and reviews.

Paperback ISBN: 978-0-578-26233-8
Hardback ISBN: 978-0-578-26237-6

π = 3 Books

Library of Congress Control Number: 2022904632

New and revised text © 2022 by James N. Akins, Jr. Certificate of Registration Number TX 9-092-173 and TX 9-121-305 Cover Illustration by Victor Guiza © 2022 by James N. Akins, Jr.

Artwork and text - © 2018 by James N. Akins, Jr.
Certificate of Registration Number TXu 2-130-773

2 D Artwork, MC2 Defined - © 2018 by James N. Akins, Jr.
Certificate of Registration Number VAu 1-398-340

PRINTED IN THE UNITED STATES OF AMERICA

Table of Content:

Chapter 1	The Message in a Bottle: Norma, I Now Know..................	2
Chapter 2	Norma, I now know what happens to Mass if it Exceeds the Speed of Light	5
Chapter 3	Norma, I now know, written by James N. Akins, Jr.	8
Chapter 4	The Story of Our Life, based on a True Life., to be written by us.	26
Chapter 5	The Story of Our Life, based on a True Life., God's Fulfilled Promise of the Truth...	30
Chapter 6	The Story of Our Life, Based on a (1) True Life. Truth is returned to Jerusalem. All mankind shall prepare for the return of Our Lord Jesus..	34
Chapter 7	Final Testament: Rise of the Holy Spirit	59

Foreword by Norma Ramos Quintero:

James N. Akins, Jr. has traveled and worked in more than 40 countries all over the world with the United States government. A man well-traveled certainly has a lot to say, and in his story, *The Message in a Bottle: Norma, I Now Know*, James captures his true-life close to death experience and vividly lets the reader in on some flashbacks and stream of conscience moments that will keep you reading till the last word.

I dedicate this book to God, Christ, the Holy Spirit Mary, Mother of Jesus and my Angels that saved my life with their words of truth and support: Dr. Ann Loudon Severance, Norma Ramos Quintero, Dr. Stephen E. Winston, MD and Jeffery J. Pappas, MSW, LICSW, PIP.

Until death did we part on April 22, 2020 my Angle Dr. Ann Loudon Severance, PhD. God sent you to me to help Him prepare the world for His Son's return on the day of the 120 th Feast of the Jubilee, December 22, 2022, but you knew that 30 years ago. You knew I was not living up to my potential and you could not go where I would be going for God. The world owes you their love for your unwavering faith in God's promises. Please forgive me for ever making you unhappy. Keep on dancing and say hello to Ruth and my father for me.

THE STORY OF OUR LIFE, Based on a True Story –

THE MESSAGE IN A BOTTLE - NORMA, I NOW KNOW

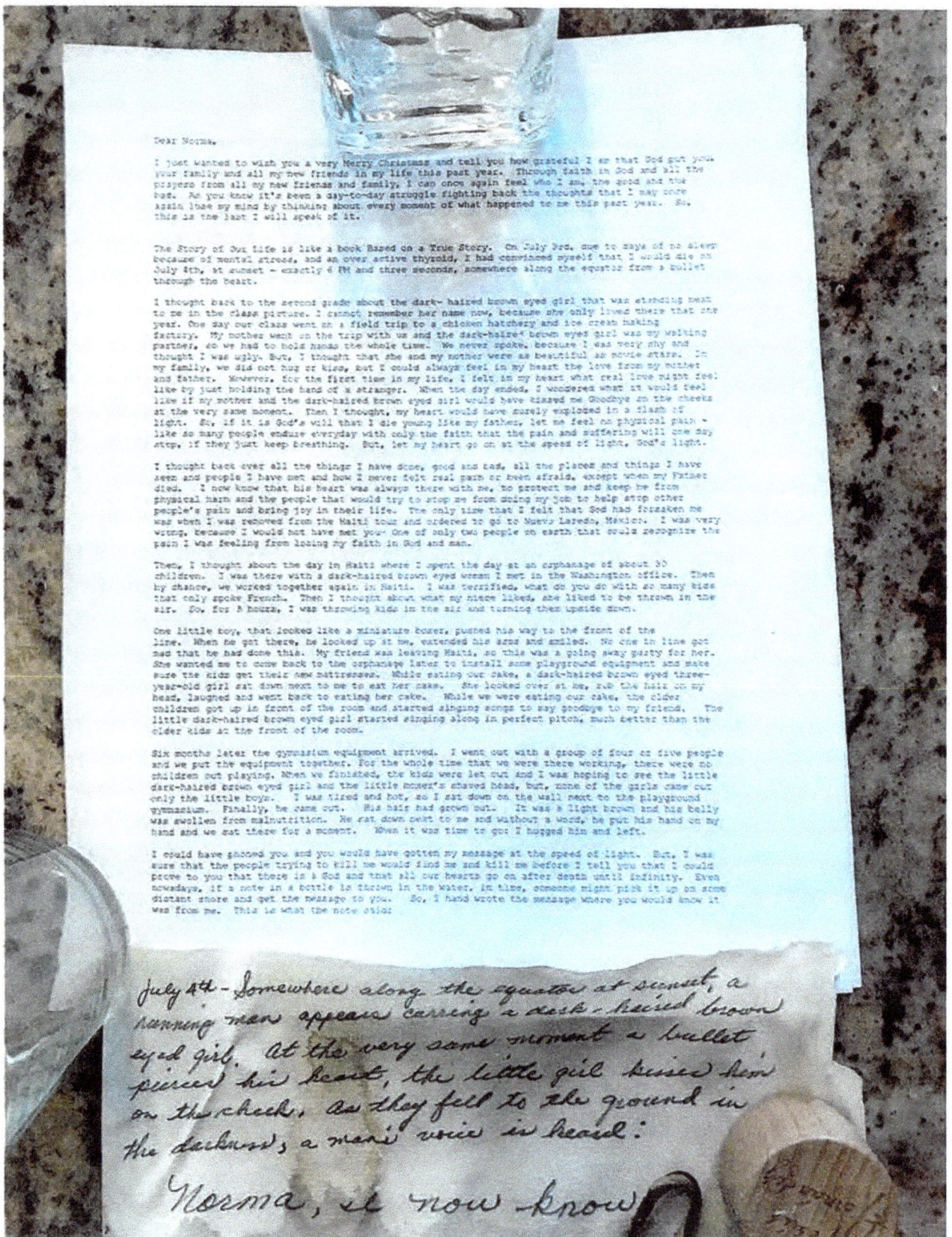

Chapter 1

THE MESSAGE IN A BOTTLE:

NORMA, I NOW KNOW

RADIO ANNOUNCER–Happy New Years. I have a special request to read a note found on a Gulf of Mexico beach in a tequila bottle.

Dear Norma,

I just wanted to wish you a very Merry Christmas and tell you how grateful I am that God put you, your family and all my new friends in my life this past year. Through faith in God and all the prayers from all my new friends and family, I can once again feel who I am, the good and the bad. As you know it's been a day-to-day struggle fighting back the thoughts that I may once again lose my mind by thinking about every moment of what happened to me this past year. So, this is the last I will speak of it.

The Story of Our Life is like a book Based on a True Story. On July 3rd, due to days of no sleep because of mental stress, and an over active thyroid, I had convinced myself that I would die on July 4th, at sunset–exactly 6 PM three seconds, somewhere along the equator from a bullet through the heart.

I thought back to the second grade about the dark-haired brown eyed girl that was standing next to me in the class picture. I cannot remember her name now, because she only lived there that one year. One day our class went on a field trip to a chicken hatchery and ice cream making factory. My mother went on the trip with us and the dark-haired brown eyed girl was my walking partner, so we had to hold hands the whole time. We never spoke, because I was very shy and thought I was ugly. But, I thought that she and my mother were as beautiful as movie stars. In my family, we did not hug or kiss, but I could always feel in my heart the love from my mother and father. However, for the first time in my life, I felt in my heart what real love might feel like by just holding the hand of a stranger. When the day ended, I wondered what it would feel like if my mother and the

The Story of Our Life, based on a True Life.

dark-haired brown eyed girl would have kissed me Goodbye on the cheeks at the very same moment. Then I thought, my heart would have surely exploded in a flash of light. So, if it is God's will that I die young like my father, let me feel no physical pain–like so many people endure everyday with only the faith that the pain and suffering will one day stop, if they just keep breathing. But, let my heart go on at the speed of light, God's light. I thought back over all the things I have done, good and bad, all the places and things I have seen and people I have met and how I never felt real pain or been afraid, except when my Father died. I now know that his heart was always there with me, to protect me and keep me from physical harm and the people that would try to stop me from doing my job to help stop other people's pain and bring joy in their life. The only time that I felt that God had forsaken me was when I was removed from the Haiti tour and ordered to go to Nuevo Laredo, Mexico. I was very wrong, because I would not have met you– One of only two people on earth that recognize the pain I was feeling from losing my faith in God and man.

Then, I thought about the day in Haiti where I spent the day at an orphanage of about 30 children. I was there with a dark-haired brown eyed woman I met in the Washington office. Then by chance, we worked together again in Haiti. I was terrified, what do you do with so many kids that only spoke French. Then I thought about what my niece liked, she liked to be thrown in the air. So, for 3 hours, I was throwing kids in the air and turning them upside down.

One little boy, that looked like a miniature boxer, pushed his way to the front of the line. When he got there, he looked up at me, extended his arms and smiled. No one in line got mad that he had done this. My friend was leaving Haiti, so this was a going away party for her. She wanted me to come back to the orphanage later to install some playground equipment and make sure the kids get their new mattresses. While eating our cake, a dark-haired brown eyed three-year-old girl sat down next to me to eat her cake. She looked over at me, rub the hair on my head, laughed and went back to eating her cake. While we were eating our cake, the older children got up in front of the room and started singing songs to say goodbye to my friend. The little dark-haired brown eyed girl started singing along in perfect pitch, much better than the older kids at the front of the room.

Six months later the gymnasium equipment arrived. I went out with a group of four or five people and we put the equipment together. For the whole time that we were there working, there were no children out playing. When we finished, the kids were let out and I was hoping to see the little dark-haired brown eyed girl and the little boxer's shaved head, but, none of the girls came out only the little boys. I was tired and hot, so I sat down on the wall next to the gymnasium. Finally, he came out. His hair had grown out. It was a light brown and his belly was swollen from malnutrition. He sat down next to me and without a word, he put his hand on my hand and we sat there for a moment. When it was time to go: I hugged him and left.

I could have phoned you and you would have gotten my message at the speed of light. But, I was sure that the people trying to kill me would find me and kill me before I tell you that I could prove to you that there is a God and that all our hearts go on after death until infinity. Even nowadays, if a note in a bottle is thrown in the water, in time, someone might pick it up on some distant shore and get the message to you. So, I hand wrote the message where you would know it was from me. This is what the note said:

July 4th, somewhere along the equator at sunset, a running man appears carrying a dark-haired brown eyed girl. At the very same moment a bullet pierces his heart, the little girl kisses him on the cheek. As they fell to the ground in darkness, a man's voice is heard: Norma, I now Know

> RADIO ANNOUNCER–This note reminds me that my brother once accidently drank a full bottle of tequila toasting colleagues, alive and dead, and ended up imagining he could exceed the speed of light, at least in the first quarter mile.
>
> Well, this song goes out to all our Brothers and Sisters that have put themselves in harm's way to make our life better, and will not make it home this New Year. On a personal note, I want to thank God and his sweet Angels for bringing my Bubba home safe this past year.
>
> Our next song is from Norma, our favorite beautiful Dark-haired Brown Eyed girl singing her latest hit: *My Heart will go on*.

Song–Every night in my dreams, I see you, I feel you, that is how I know you go on.

Chapter 2

NORMA, I NOW KNOW what happens to Mass if it Exceeds the Speed of Light.

In the YouTube video, you sent me; There Is A God - Lee Ann Womack, there is a picture of Dr. Albert Einstein [Footnote No. 1]. When I viewed the video, I had not slept for week and I was thinking about nothing but, how to explain to you; "What happens to mass if it exceeds the speed of light?" I thought, how could God let him tell the world the scientific formula for the Atom Bomb and how could he ever smile again after he saw the how many humans died and the subsequent "Cold War" that followed [Dr.Albert Einstein died in1955 the year after I was born.] The whole world believed Dr. Albert Einstein's formula "$E=mc^2$" was the key to how the Atom Bomb works. Before they set off the first Atom Bomb at White Sands, New Mexico [I worked a few miles from there in Alamogordo, NM.] the scientist thought the Atom Bomb would cause a chain reaction in the atmosphere and destroy earth. Well, the world did not blow up, but more than 115 thousand humans were killed or deformed from the two bombs that were dropped on Japan [Fortunately, millions of humans were saved by convincing the Emperor of Japan to stop the War.]

So, I broke Dr. Albert Einstein's formula down by defining each of the letters and making a sentence out of them to see if the formula made any since:

E [ALL THE ENERGY IN THE UNIVERSE] = [EQUALS] m [ALL THE MASS IN THE UNIVERSE] (In physics, mass is a property of a physical body, is often seen as spherical because it is such a fundamental quantity that it is hard to define in terms of something else.) times c^2 [c = CONSTANT SPEED OF LIGHT times c = CONSTANT SPEED OF LIGHT.]

The first of two atom bombs that exploded over Japan was called "Little Boy," its design used the gun method to explosively push 38.5 (85 pounds) kilograms hollow cylinder of sub-critical mass of uranium-235 1.100 meters (42 inches), by means of four cylindrical silk bags of cordite (Burn rate 1,000 meter second), into a "target" 25.6 (56 pounds) kilogram solid target cylinder together into supercritical mass, initiating a nuclear chain reaction.

The Story of Our Life, based on a True Life.

To this date, physicist express "$E = mc^2$ " as the dynamic equation for calculating FORCE to measure the ENERGY of the atom bombs. So, now the sentence reads: E [FORCE = JOULES (38,500,000-kilograms times meter² per second²] = [EQUALS] m [38.5 kilograms mass] times c^2 = a (acceleration) [1,000 meters per second times 1,000 meters per second or 1,000,000-meter² per second².] 38,500,000 Joules is a far cry less than 38.5 kilograms mass times c^2 (Constant Speed of Light² = 89,875,517,873,681,800-meter²/second²) or 3,460,207,438,000,000,000 Joules.

As anyone can see $E = mc^2$ is not the same Dynamic equation as $F = ma$, because no explosives on earth can propel a mass into another mass or implode one mass into another mass at 89,875,517,873,681,800-meter²/second².

Mass has two states of existence; Stop or Static and Moving or Dynamic. If one simply changes the Dynamic Force and $E=mc^2$ equations to Static equations by removing the movement of acceleration and Constant Speed of Light² the Force equation would read; mass (kilograms) = mass (kilograms) and Einstein's equation would read; ENERGY = MASS (E=M), therefore, it is a false assumption that the two Dynamic equations are equal.

Did Dr. Einstein mislead the scientific community by not openly addressing the meaning of his equation as I just did? I do not think so, because he used a small letter "m" in his formula to represent a quantity of mass less than "ALL THE MASS IN THE UNIVERSE". I believe any high school math student knows the difference between a Static equation and a Dynamic equation and how to check one's work. Norma, I now know what $E = Mc^2$ (Big letter "M") means.

"If you knew the magnificence of the three, six and nine, you would have a key to the universe." – Nikola Tesla

Even if he did not know, the formula $E=Mc^2$ clearly defined the magnificence of 3, 6, and 9. His formula could have also been written: (E=M) [rotating sphere of magnetic spheres] = ½ (E=M) (c) · ½ (E=M) (c). Also, I have mathematical proof that Pi = 3 not 3.14.

The information in Chapter 3, may lead to more efficient power generation and electric motors, faster computers, more efficient computer operating systems and memory storage, Big Bang Law, String Law, and Law of General Relativity, cure for cancer, define Black Holes, fine more Goldilocks planets and how to communicate with other planets.

Norma, I now know it is God's law that no mass may exceed the speed of light and Dr. Albert Einstein could smile and be happy, because of his faith in God's love for him and the lives he did help save during such a horrific time in human history.

NEWS! NEWS! NEWS! NEWS! NEWS! NEWS!

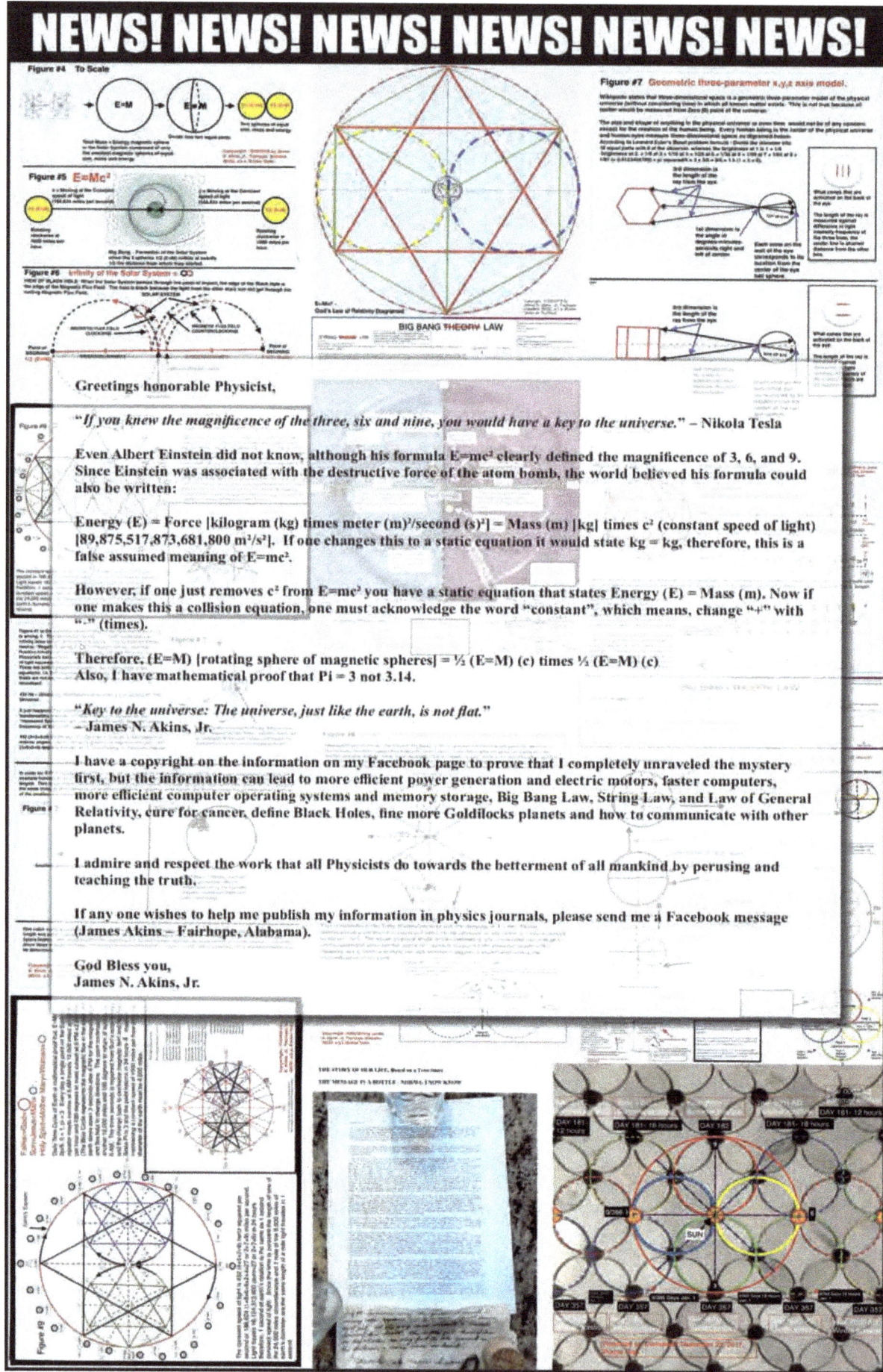

Greetings honorable Physicist,

"If you knew the magnificence of the three, six and nine, you would have a key to the universe." – Nikola Tesla

Even Albert Einstein did not know, although his formula $E=mc^2$ clearly defined the magnificence of 3, 6, and 9. Since Einstein was associated with the destructive force of the atom bomb, the world believed his formula could also be written:

Energy (E) = Force [kilogram (kg) times meter $(m)^2$/second $(s)^2$] = Mass (m) [kg] times c^2 (constant speed of light) [89,875,517,873,681,800 m^2/s^2]. If one changes this to a static equation it would state kg = kg, therefore, this is a false assumed meaning of $E=mc^2$.

However, if one just removes c^2 from $E=mc^2$ you have a static equation that states Energy (E) = Mass (m). Now if one makes this a collision equation, one must acknowledge the word "constant", which means, change "+" with "." (times).

Therefore, (E=M) [rotating sphere of magnetic spheres] = ½ (E=M) (c) times ½ (E=M) (c).
Also, I have mathematical proof that Pi = 3 not 3.14.

"Key to the universe: The universe, just like the earth, is not flat."
– James N. Akins, Jr.

I have a copyright on the information on my Facebook page to prove that I completely unraveled the mystery first, but the information can lead to more efficient power generation and electric motors, faster computers, more efficient computer operating systems and memory storage, Big Bang Law, String Law, and Law of General Relativity, cure for cancer, define Black Holes, fine more Goldilocks planets and how to communicate with other planets.

I admire and respect the work that all Physicists do towards the betterment of all mankind by perusing and teaching the truth.

If any one wishes to help me publish my information in physics journals, please send me a Facebook message (James Akins – Fairhope, Alabama).

God Bless you,
James N. Akins, Jr.

Chapter 3

I now know..., written by James N. Akins, Jr.

Figure #1 is the present understanding of the meaning of $E=mc^2$, it is wrong: 1. The symbol for infinity ∞ and the definition of infinity (also known as Forever) is wrong. The symbol for Infinity means; "Negative Infinity = ∞ = Positive Infinity", whereas, Positive Infinity minus Negative Infinity = Pi (Not Forever). 2. Physicists believe $E=mc^2$ (Energy=mass times the constant speed of light squared) is equal to F=ma (Force=mass times acceleration). These are both Dynamics equations. By making them Static equations, i.e, E(Energy)=m(Mass) is equal to (F)Force = m (mass) these are not equal equations, because Force is just Mass without movement.

432 Hz – Unlocking The Magnificence of the 3 6 9: The Key To The Universe.

It just happens that the constant speed of light is 432 cubic handbreadths squared, or 186,624 miles per second, not the "measured Speed of Light" and verified by Music Standard tuning frequency of Verdi's 'A' = 432 Hz (also, the frequency of red light.)

432 (4+3+2=9) divided by 12 = 36 (3+6= 9) A circle is 360 degrees in interior angles. Divide 360 by 10 parts is 36 degrees each part. 360 (3+6+0=9) degrees divided by 12 (1+2=3) hours is 30 (3+0=3) degrees.

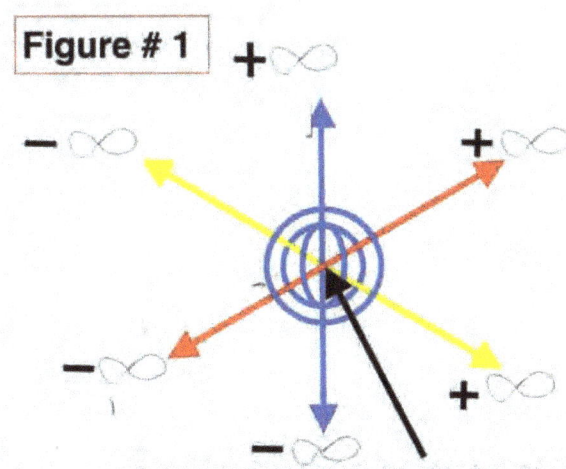

Present understanding of the Big Bang theory is $E=mc^2$ caused the Big Bang, whereas, there was an explosion that cooled down and formed the Universe into infinity in all directions from 0.

In order for E=M or Energy = Mass they must be the same thing and they must be definite in size not indefinite or unlimited. For example humans believe that in the Universe, it is possible to have something smaller than the smallest and larger than the largest. This is not true in our Solar system. The Universe outside our Solar System is made up of their own $E=Mc^2$'s. Energy is the same thing as Mass because the smallest particle is a magnetic solid sphere that started out as a large solid sphere made up of the smallest magnetic spheres.

Not to scale:

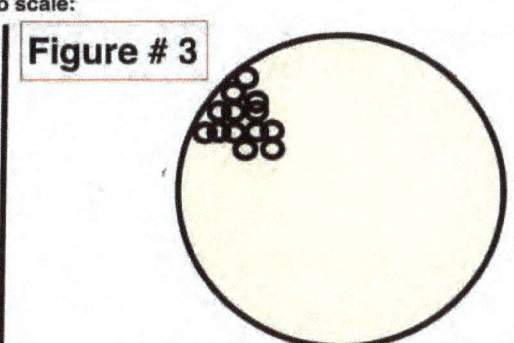

Figure # 2

o

Smallest particle magnetic sphere

Figure # 3

Total Mass = Energy magnetic sphere in the Solar System composed of only the smallest magnetic spheres of equal size, mass and energy.

One cubic centimeter of water is one gram (Mass). The measurement unit is not significant, only that all sides are equal in length and water's mass is exactly equal to 1. Place 1 gram of water in a vacuum to form a sphere, like in the International Space Station. The actual length of the diameter is 1 measuring unit according to the geometry of a circle and not 1 centimeter. Since Mass is equal to Energy, the energy maybe be measured in 1 gram of water and the diameter of the smallest spheres may be determined.

Copywright, 10/09/2018 by James N. Akins, Jr., Fairhope, Alabama 36532. a.k.a. Bubba Twain

Figure #4 To Scale

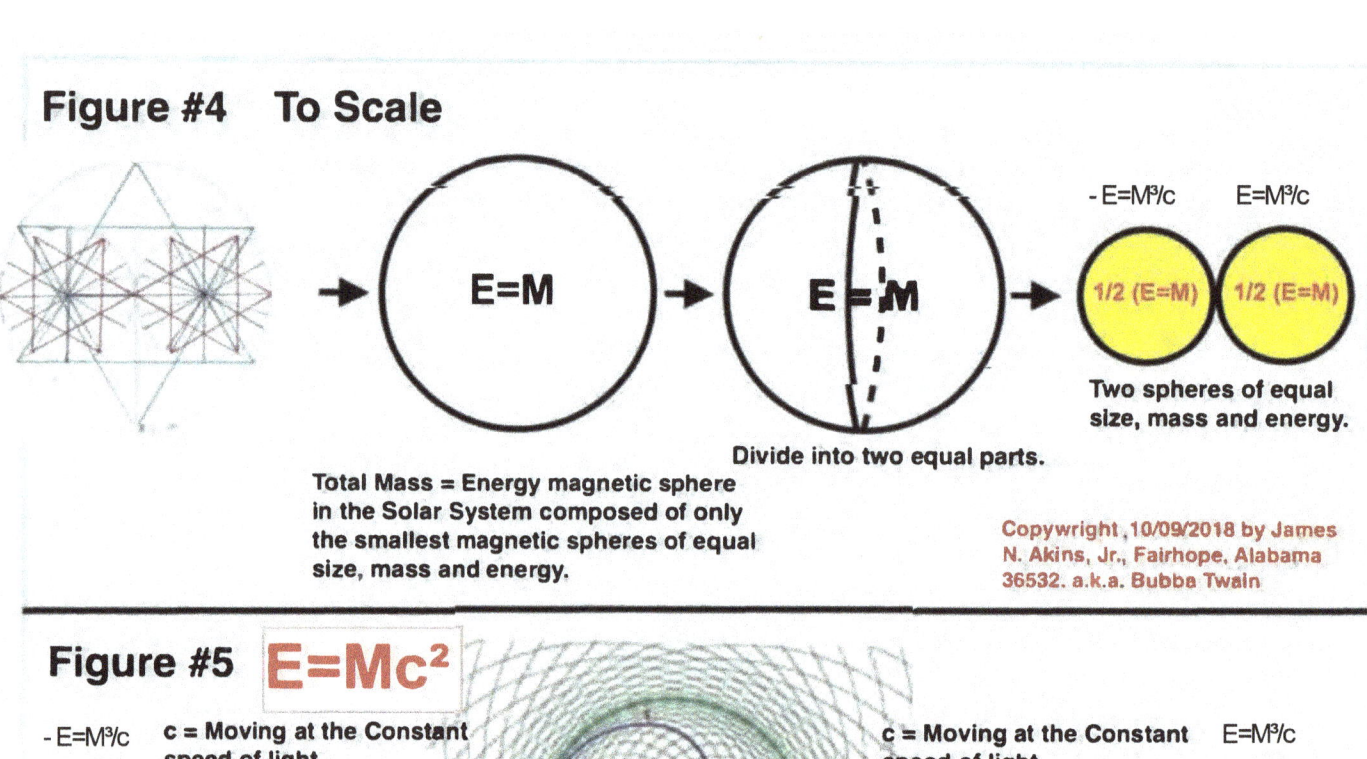

Total Mass = Energy magnetic sphere in the Solar System composed of only the smallest magnetic spheres of equal size, mass and energy.

Divide into two equal parts.

Two spheres of equal size, mass and energy.

Copyright, 10/09/2018 by James N. Akins, Jr., Fairhope, Alabama 36532. a.k.a. Bubba Twain

Figure #5 $E=Mc^2$

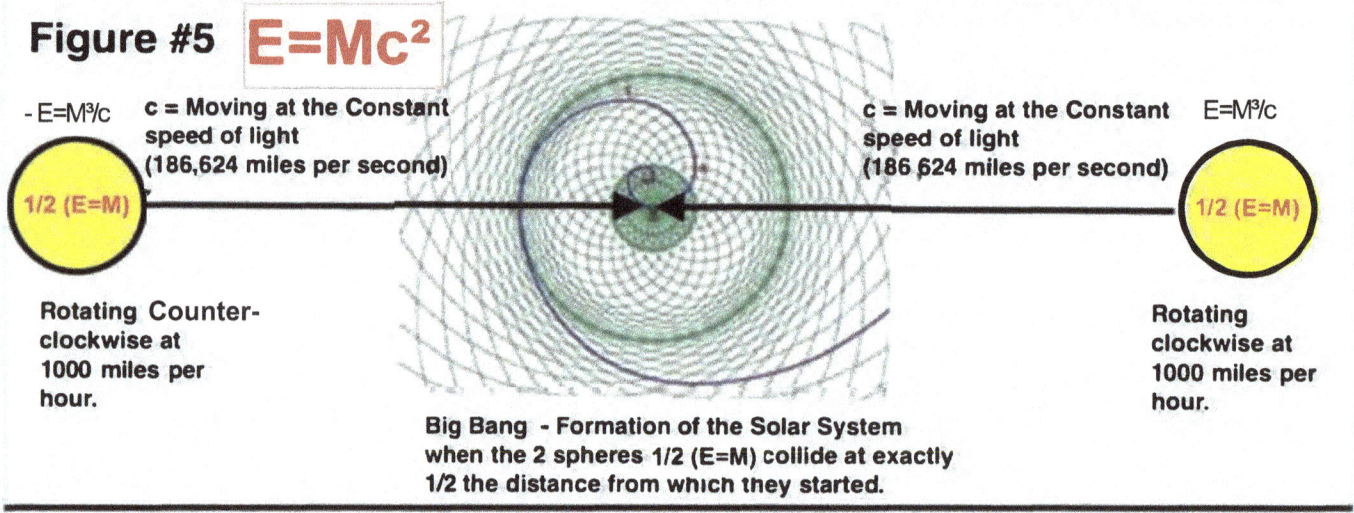

$-E=M^3/c$

c = Moving at the Constant speed of light (186,624 miles per second)

1/2 (E=M)

Rotating Counter-clockwise at 1000 miles per hour.

c = Moving at the Constant speed of light (186,624 miles per second)

$E=M^3/c$

1/2 (E=M)

Rotating clockwise at 1000 miles per hour.

Big Bang - Formation of the Solar System when the 2 spheres 1/2 (E=M) collide at exactly 1/2 the distance from which they started.

Figure #6 Infinity of the Solar System = ∞

VIEW OF BLACK HOLE- When the Solar System passes through the point of impact, the edge of the Black Hole is the edge of the Magnetic Flux Field. The hole is black because the light from the other stars can not get through the trailing Magnetic Flux Field.

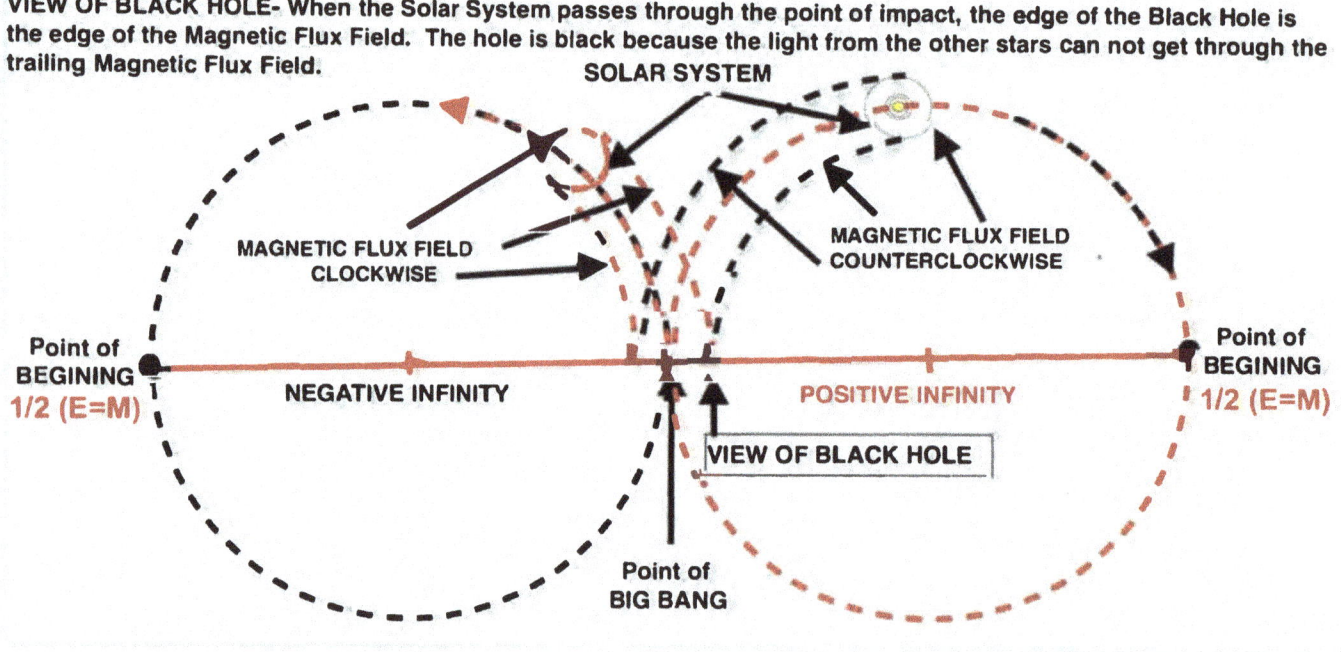

Figure #7 Geometric three-parameter x,y,z axis model.

Wikiquote states that three-dimensional space is a geometric three-parameter model of the physical universe (without considering time) in which all known matter exists. This is not true because all matter would be measured from Zero (0) point of the universe.

The size and shape of anything in the physical universe or even time would not be of any concern except for the creation of the human being. Every human being is the center of the physical universe and human eyes measure three-dimensional space as digramed below.
According to Leonard Euler's Basel problem formula : Divide the diameter into 10 equal parts with 0 at the observer, whereas the brightness at 1 is 1 + 1/4 brightness at 2, + 1/9 at 3 + 1/16 at 4 + 1/25 at 5 + 1/36 at 6 + 1/49 at 7 + 1/64 at 8 + 1/81 (= 0.0123456789) = pi squared/6 = 3 x 3/6 = 9/6 = 1.5 (1 + 5 = 6).

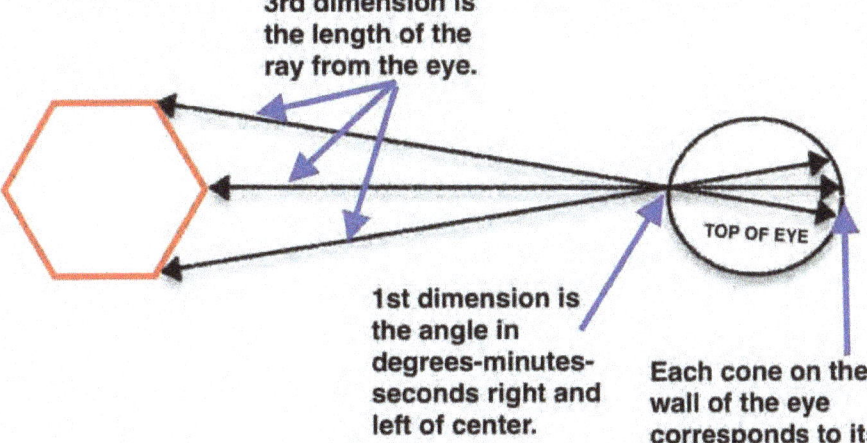

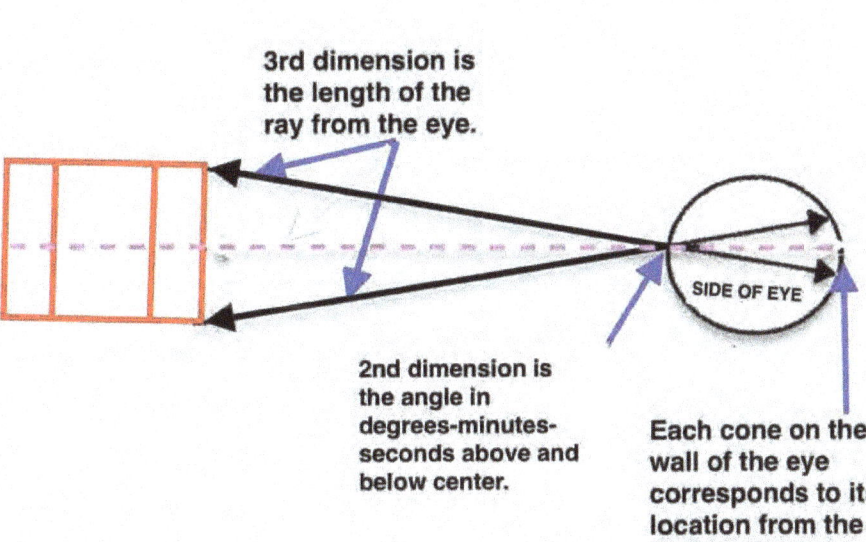

Copywright, 10/09/2018 by James N. Akins, Jr., Fairhope, Alabama 36532. a.k.a. Bubba Twain

Figure #8 $E = Mc^2$: [1 = 0], [2Pi/6], [Pi = 3], [(+) Infinity minus (−) Infinity = Pi]

Referance Figure #6 below, The 'Point of Big Bang" is [0 = 1] because it starts at 0 and goes 360 degrees in the clockwise direction back to the same point that is now 1, then it is 0 again and goes 360 degrees in the counterclockwise direction back to the same point that is 1 again. In both cases it is 1 full revolution equaling a distance of Pi for each circle with a diameter of 1d. Disconnect at 1 = 0 point and move top 1/2 up 1d and bottom 1/2 down 1 d. Divide the 2pi circle into 6 parts as indicated by the blue lines. See diagram below.

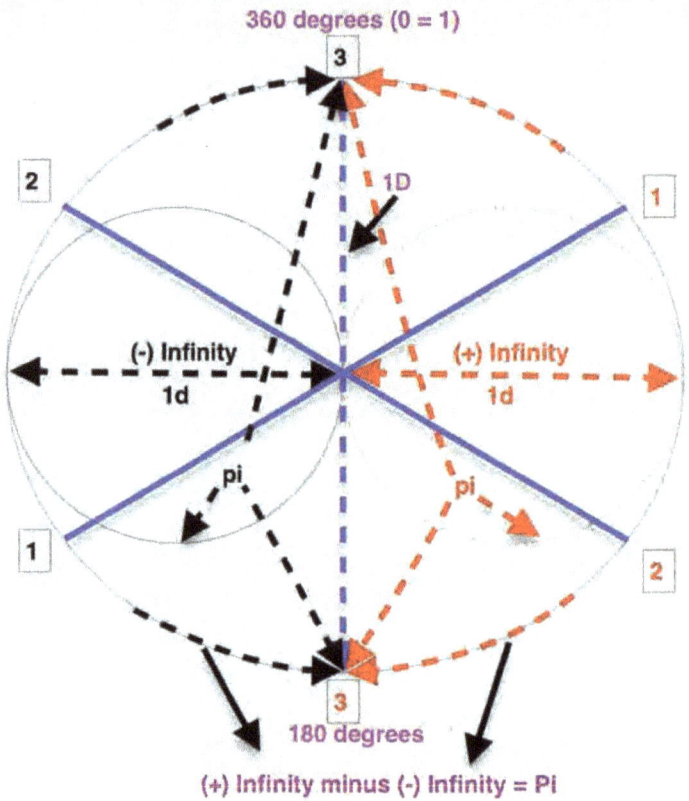

2pi/6

This is the order of the Solar System/Universe and the meaning of $E = Mc^2$. This is mathematical proof that Pi is equal to 3 not 3.14, the diameter of any circle is 1 and is always a ratio of 1 to 3. The actual physical length of the diameter of any circle does not change a circles geomeery because the radius of a circle is always 0.5 the physical length of the diameter and a circle is divided into 360 individual degree of equal length along the circumferance of the circle.

Figure #6   Infinity of the Solar System = ∞

VIEW OF BLACK HOLE- When the Solar System passes through the point of impact, the edge of the Black Hole is the edge of the Magnetic Flux Field. The hole is black because the light from the other stars can not get through the trailing Magnetic Flux Field.

Copywright, 10/09/2018 by James N. Akins, Jr., Fairhope, Alabama 36532. a.k.a. Bubba Twain

Father=God= ◯
Son=Jesus=Man= ◯
Holy Spirit=Mother Mary=Woman= ◯

Daily Time Cycle of Earth is mathmatical proof that $E=Mc^2$, $2pi/6$, $0 = 1$, $pi = 3$. Every day a single point on the Earth's equator meets sunrise at 6 AM travels 12,000 miles at 1,000 per hour and 180 degrees to meet sunset at 6 PM +3 seconds (The Blue Circle represents the magnetic flow in the earth, the earth slows down 3 seconds after 6 PM for the magnetic flow and flux field to change directions. The point continues another 12,000 miles and 180 degrees to return at sunrise at 6 AM. The three seconds is regained from Sun's solar energy and the change back to clockwise magnetic field and flux field. Since Pi is 3 and the point returns in 24 hours it must be maintaining a constant speed of 1000 miles per hour and the diameter of the earth must be 8,000 miles.

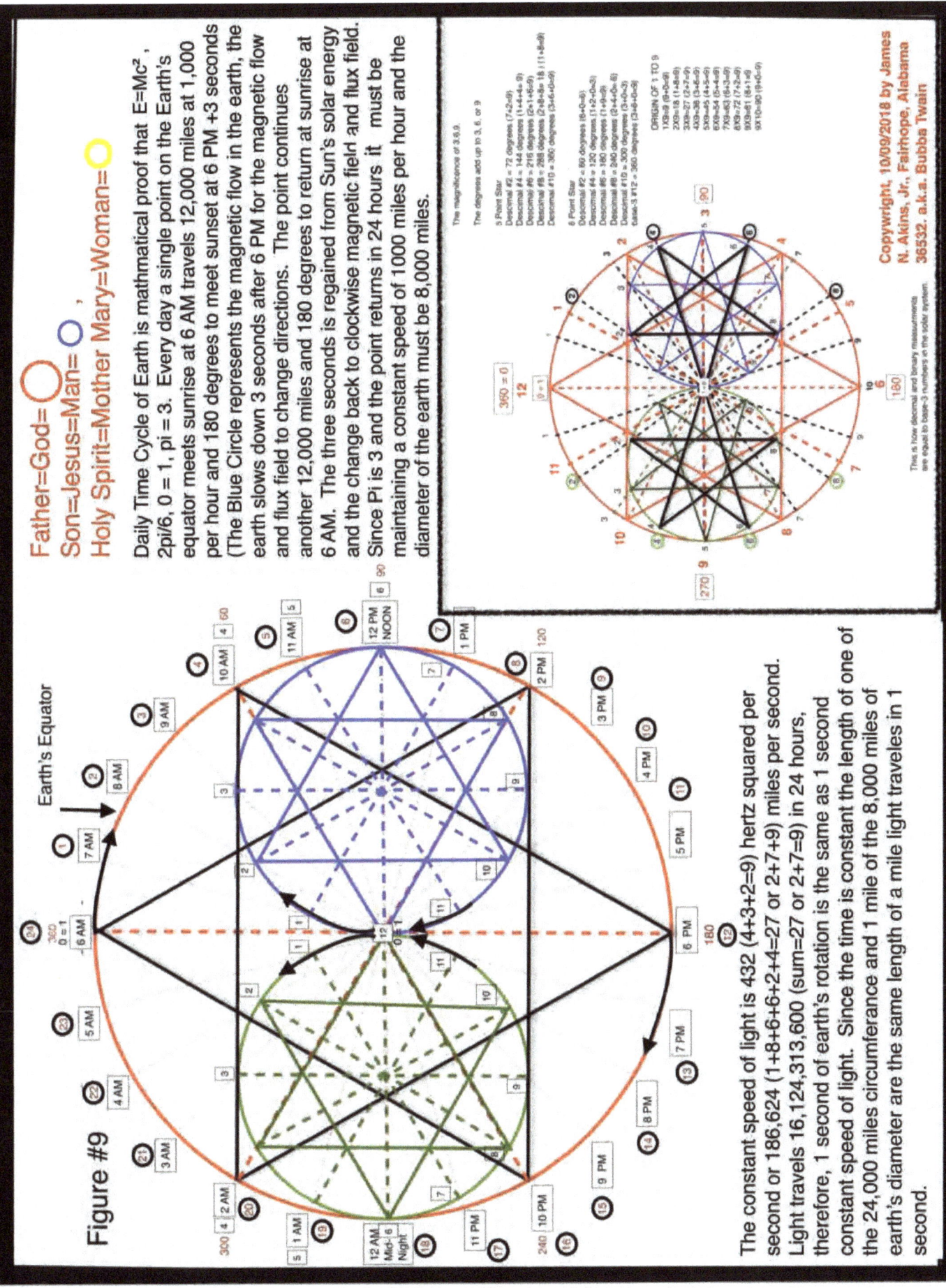

Figure #9

The magnificence of 3,6,9.

The degrees add up to 3, 6, or 9

3 Point Star
Decimal #2 = 72 degrees (7+2=9)
Decimal #4 = 144 degrees (1+4+4= 9)
Decimal #6 = 216 degrees (2+1+6=9)
Decimal #8 = 288 degrees (2+8+8= 18) (1+8=9)
Decimal #10 = 360 degrees (3+6+0=9)

6 Point Star
Decimal #2 = 60 degrees (6+0=6)
Decimal #4 = 120 degrees (1+2+0=3)
Decimal #6 = 180 degrees (1+8=9)
Decimal #8 = 240 degrees (2+4+0=6)
Decimal #10 = 300 degrees (3+0=3)
base-3 #12 = 360 degrees (3+6+0=9)

ORIGIN OF 1 TO 9
1X9=9 (9+0=9)
2X9=18 (1+8=9)
3X9=27 (2+7=9)
4X9=36 (3+6=9)
5X9=45 (4+5=9)
6X9=54 (5+4=9)
7X9=63 (6+3=9)
8X9=72 (7+2=9)
9X9=81 (8+1=9)
9X10=90 (9+0=9)

This is how decimal and binary measurments are equal to base-3 numbers in the solar system.

Copywright, 10/09/2018 by James N. Akins, Jr., Fairhope, Alabama 36532. a.k.a. Bubba Twain

The constant speed of light is 432 (4+3+2=9) hertz squared per second or 186,624 (1+8+6+6+2+4=27 or 2+7+9) miles per second. Light travels 16,124,313,600 (sum=27 or 2+7=9) in 24 hours, therefore, 1 second of earth's rotation is the same as 1 second constant speed of light. Since the time is constant the length of one of the 24,000 miles circumferance and 1 mile of the 8,000 miles of earth's diameter are the same length of a mile light traveles in 1 second.

Figure #10 $E=Mc^2$

$0 = 1$
$0 = 360$

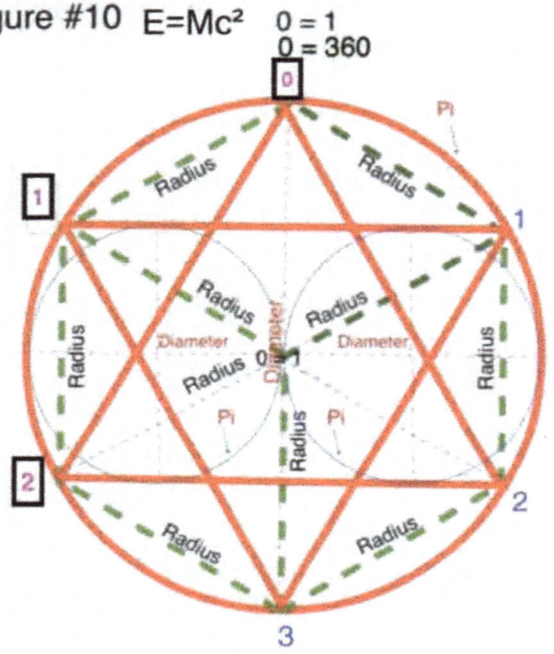

Diameter of a circle always equals 1.
Radius of a circle is always equal to 0.5.
Pi is always 3.
$0 = 1$
$0 = 1 = 360$ degrees

Circumference of a circle layed out in a straight line

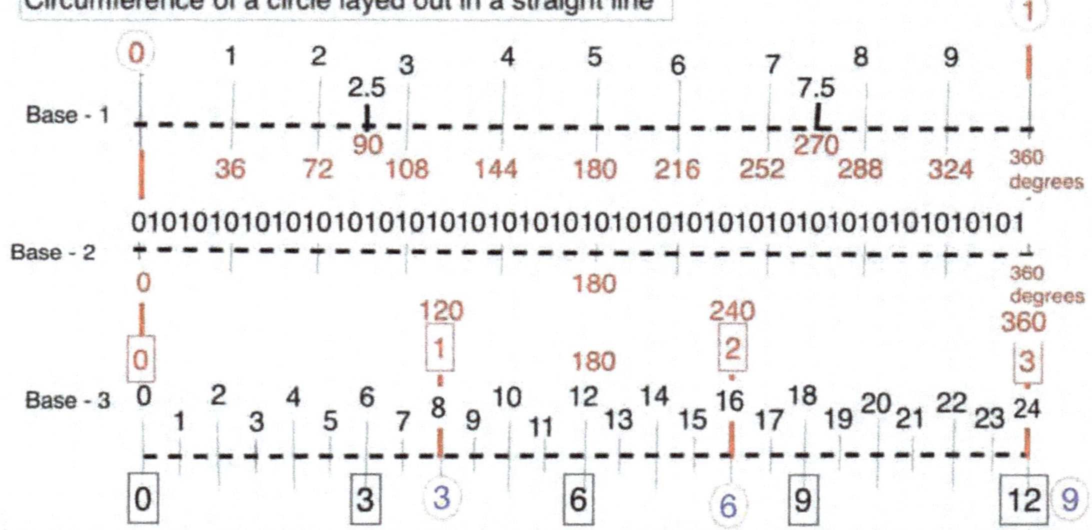

Base - 1: 0, 1, 2, 2.5, 3, 4, 5, 6, 7, 7.5, 8, 9, 1
        0, 36, 72, 90, 108, 144, 180, 216, 252, 270, 288, 324, 360 degrees

Base - 2: 0, 0101010101010101010101010101010101010101
        0, 120, 180, 240, 360 degrees
        1, 2, 3

Base - 3: 0, 1, 2, 3, 4, 5, 6, 7, 8, 9, 10, 11, 12, 13, 14, 15, 16, 17, 18, 19, 20, 21, 22, 23, 24
        180, 3, 6, 9, 12, 3

Note the green line tracing a geometric cube in two dimensions. The third dimention - Reference Figure #7, if this was a physical solid cube your eyes determine the distance of each point on the 3 surfaces by the different frequencies/brightness of green from shortest to longest distance from your eyes.

Combined with the above diagram for $E=Mc^2$, the diagram on right is mathematical proof God created the universe in 6 days (360 days around a circle and 6 days (24 Years) traveling in a spiral.) and rested on the 7th day (4 Years).

Every 4 years on leap year we add 1 day. This is not possible because the Earth would have left the Sun's gravity pull a long time ago. The earth is able to stay in orbit of the Sun by being pulled 1 day closer to the Sun then resisting its gravity to reach maximum orbit of 366 days every 8 Years.

Copyright, 10/09/2018 by James N. Akins, Jr., Fairhope, Alabama 36532. a.k.a. Bubba Twain

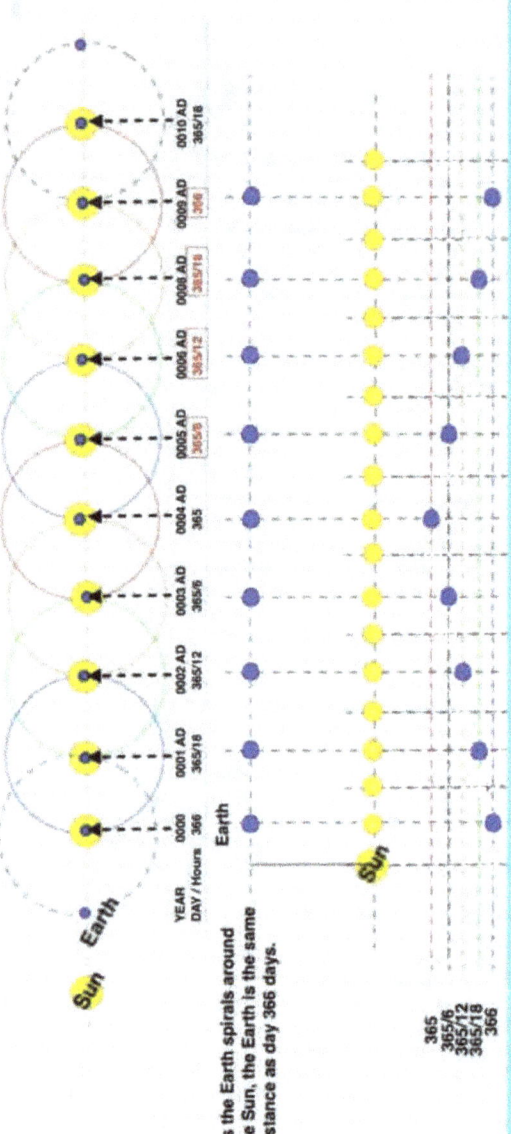

As the Earth spirals around the Sun, the Earth is the same distance as day 366 days.

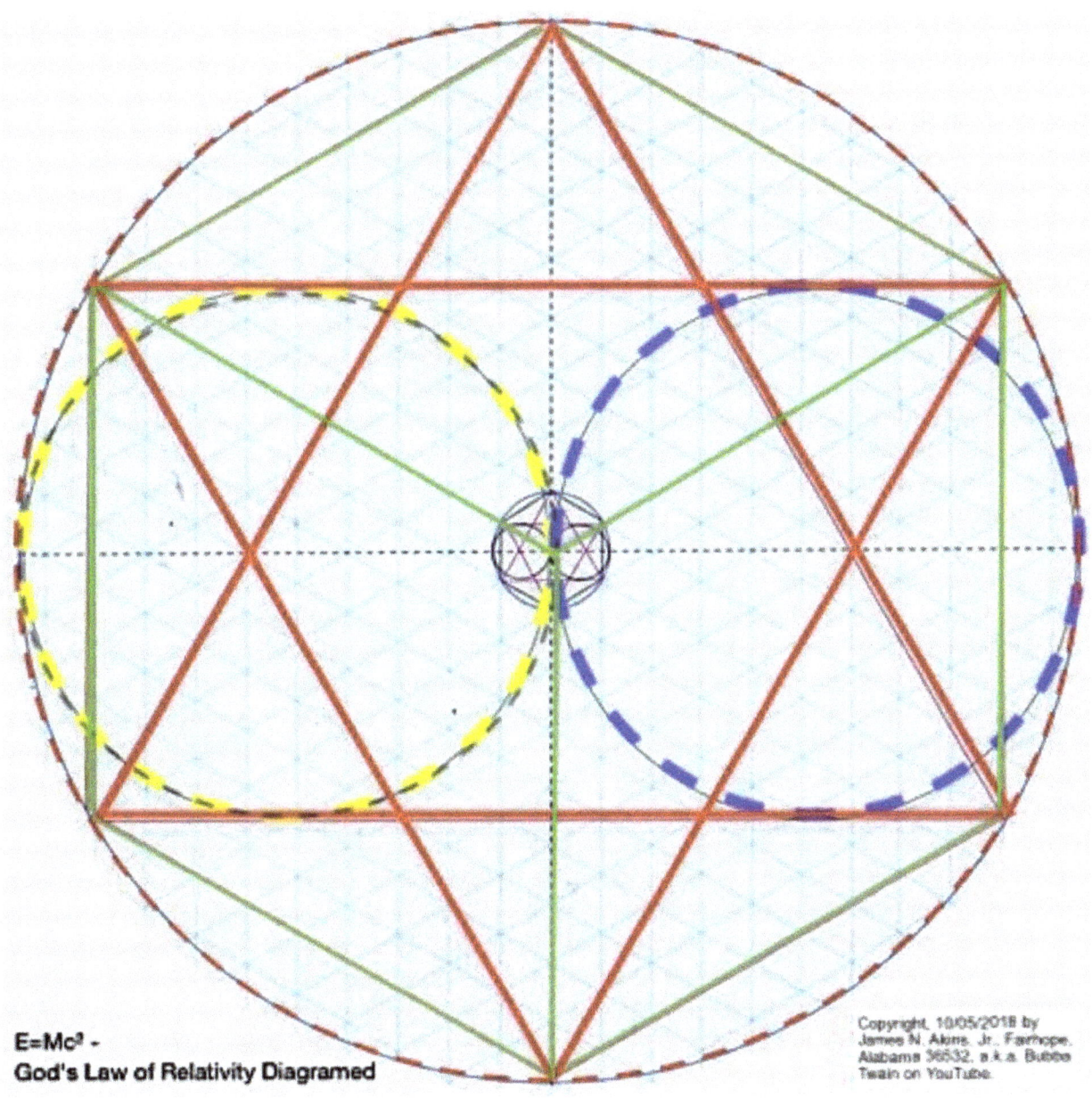

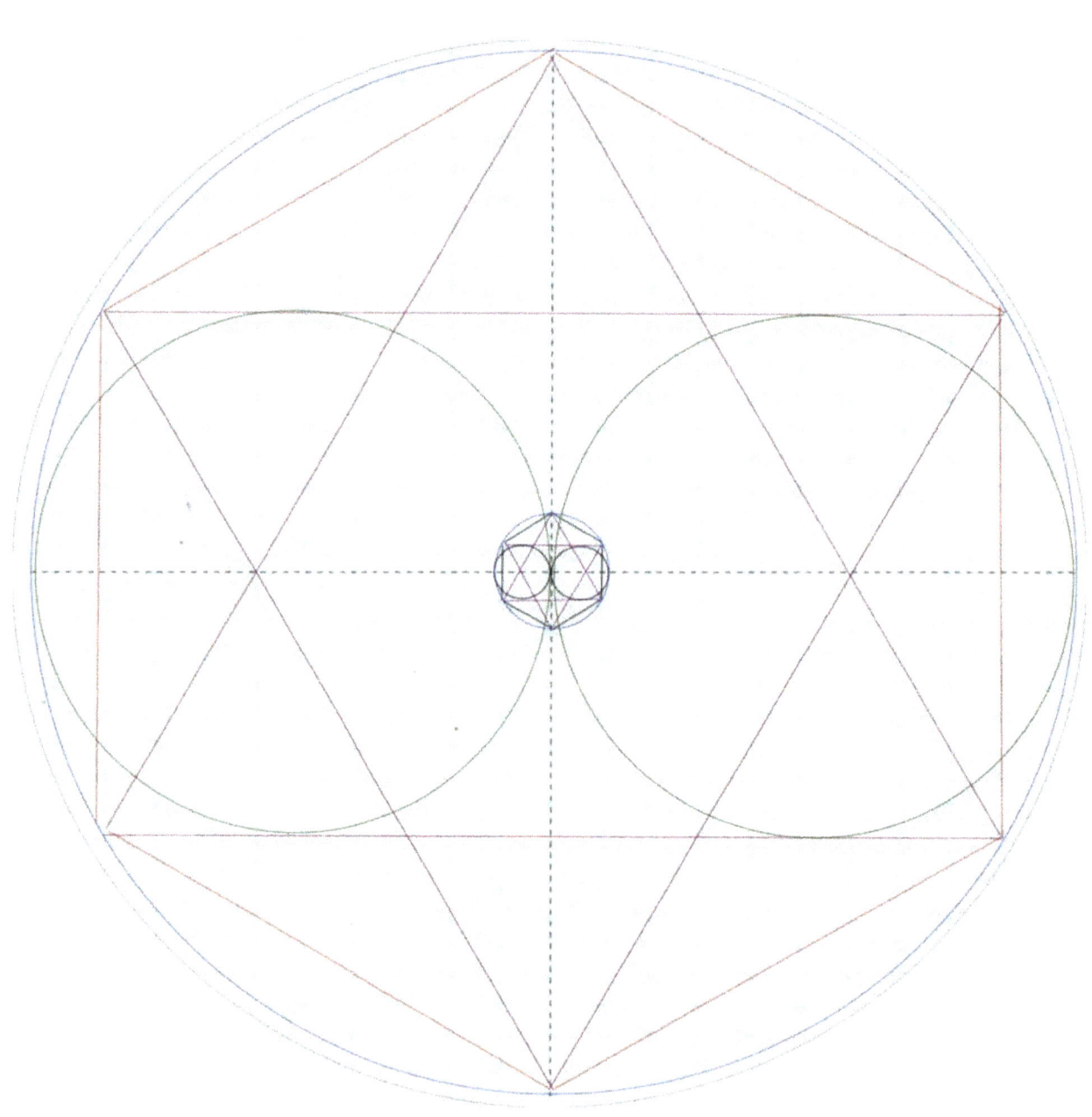

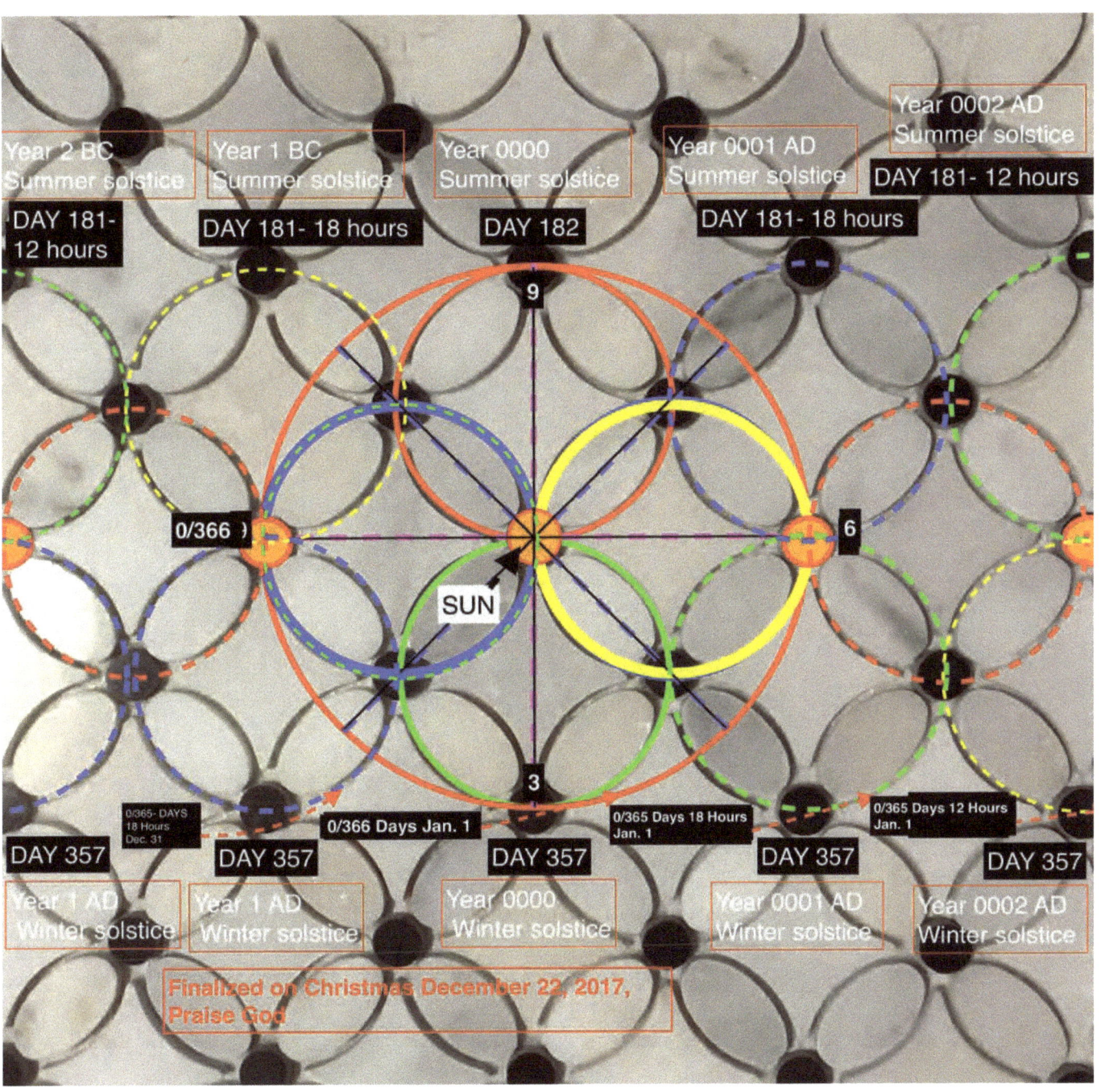

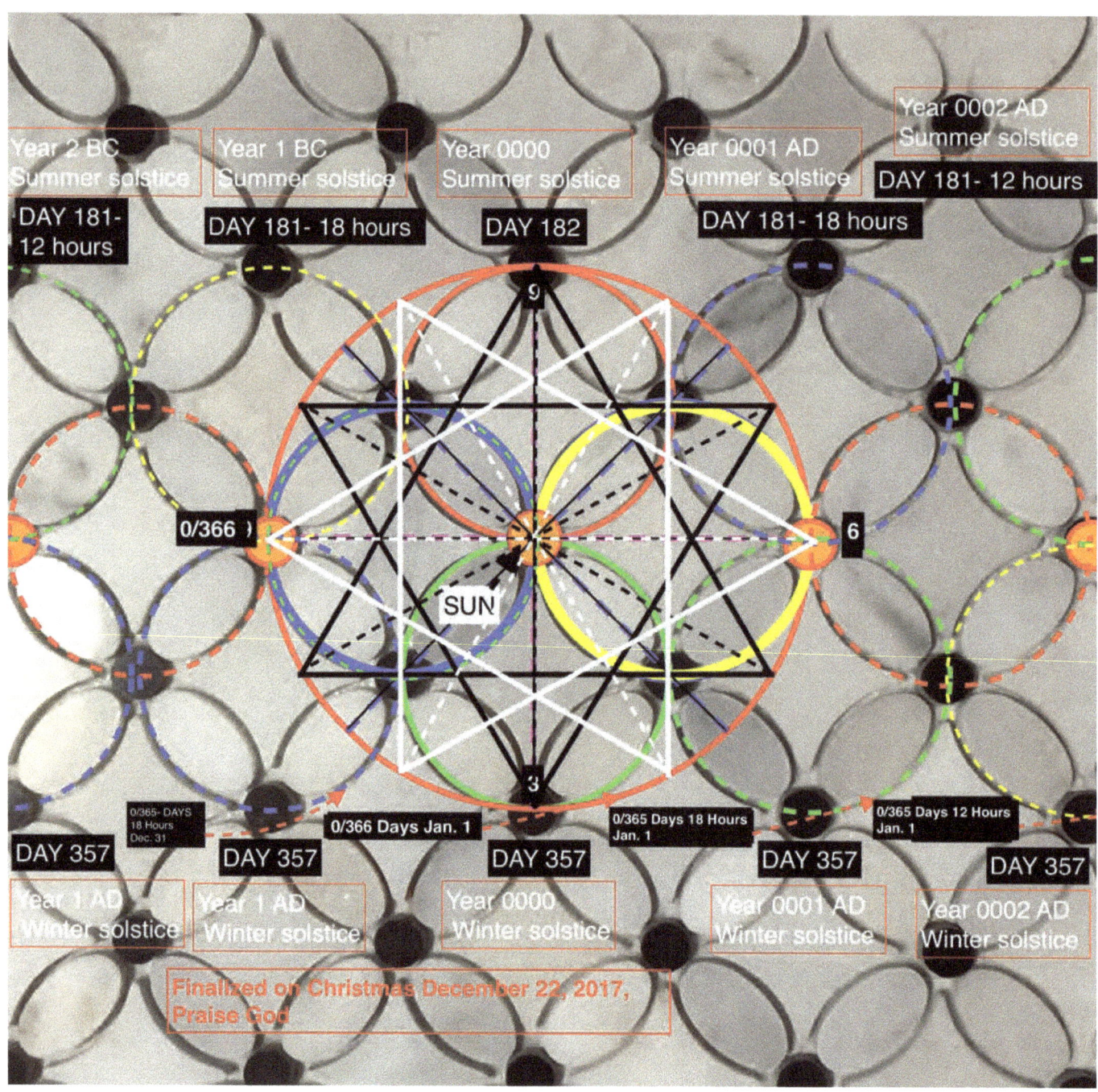

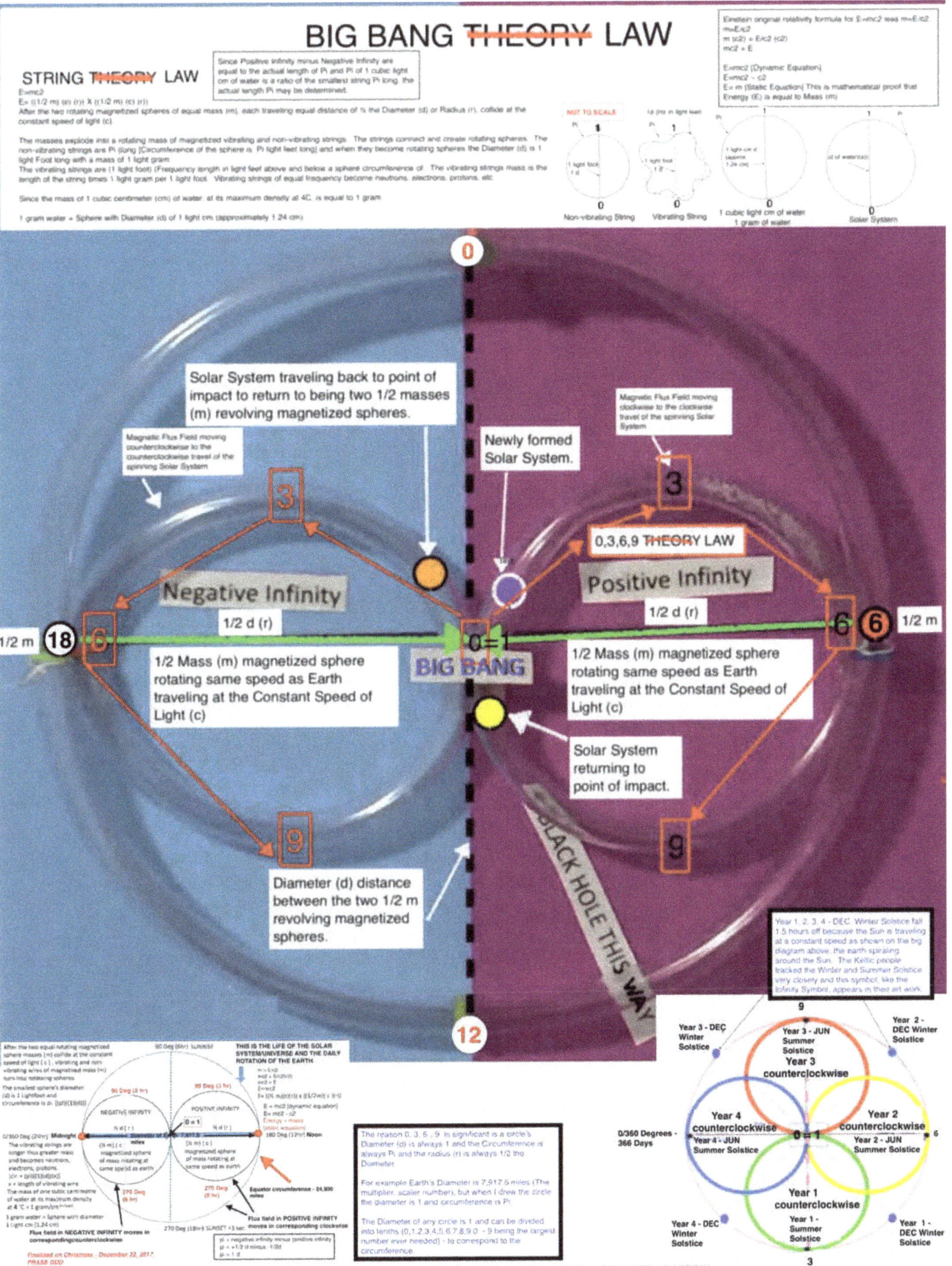

BIG BANG THEORY LAW
STRING THEORY LAW
Solar System/Universe Formed
Pi
Rotating clockwise
1/3 mass
(1)
Circumference 110
20/24
22
23
21
1/3 mass
(1)
1/3 mass
(0)
Movement through POSITIVE INFINITY and NEGATIVE INFINITY.
Solar System/Universe Process Reversed
Black Hole this way.
This is the DYNAMIC movement to the CREATION of the Solar System/Universe, therefore it took 6 days (Each day is 4 years long) to get to the final 110 Circumference.
Moment of Big Bang*
* From moment of the Big Bang until the beginning of the 6 days of the CREATION of the Universe, is 360 days each 4 years long or 1,440 years. Note: 1+4+4+0 = 9.
This circle is STATIC, therefore there are 360 Degrees inside the circle each Degree is equal to one (1) Day.
10
0
1/2 mass
(0)
1/2 mass
(1)
Pi
Rotating clockwise
Maximum Impact
5
0/20
10 Radius
10 Radius
1/2 mass
(0)
1/2 mass
(1)
15
Pi
Rotating Clockwise
Moment Before Impact
9
Year 3 - DEC Winter Solstice
Year 3 - JUN Summer Solstice
Year 3 counterclockwise
Year 2 - DEC Winter Solstice
Year 4 counterclockwise
Year 2 counterclockwise
0/360 Degrees - 366 Days
Year 4 - JUN Summer Solstice
0 = 1
Year 2 - JUN Summer Solstice
6
Year 1 counterclockwise
Year 4 - DEC Winter Solstice
Year 1 - Summer Solstice
Year 1 - DEC Winter Solstice
3

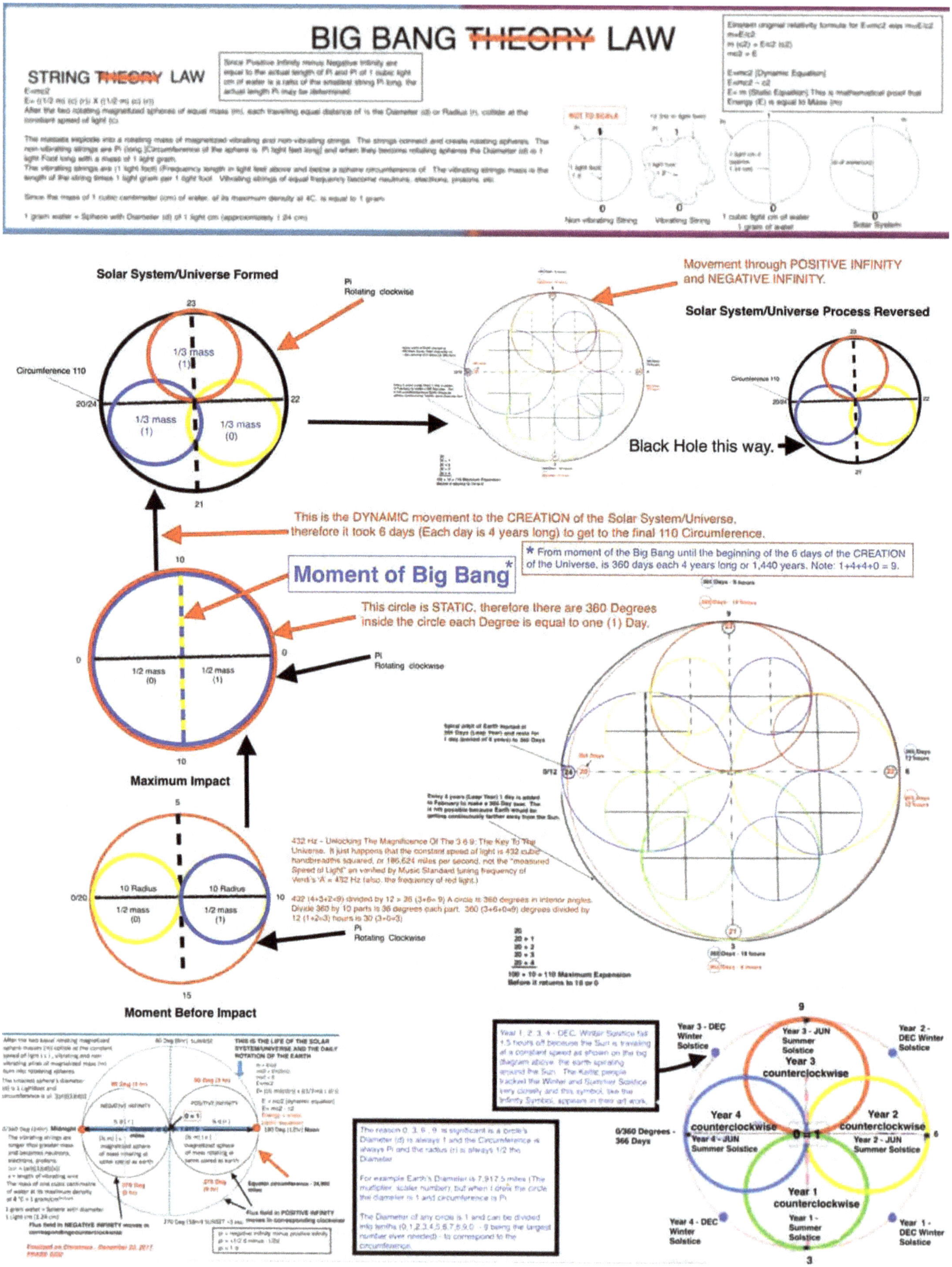

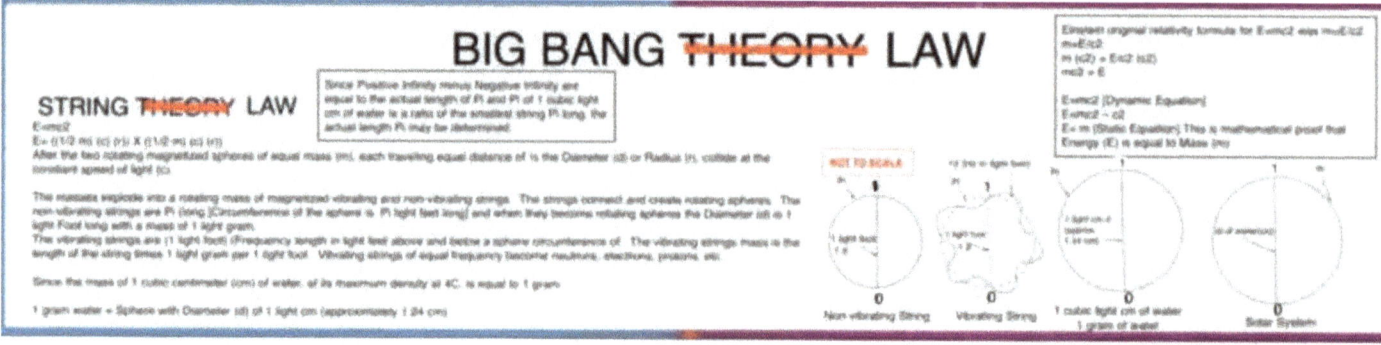

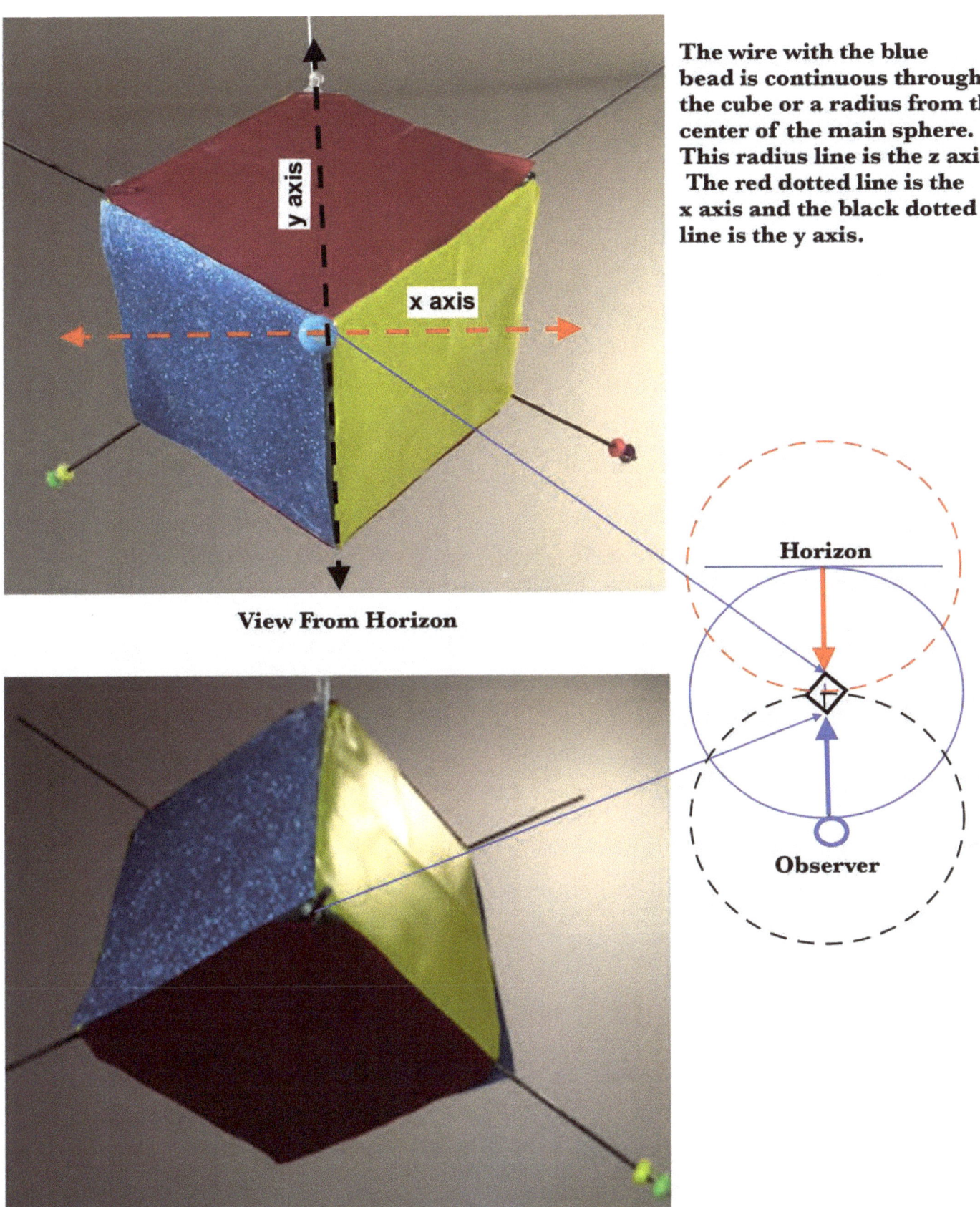

The wire with the blue bead is continuous through the cube or a radius from the center of the main sphere. This radius line is the z axis. The red dotted line is the x axis and the black dotted line is the y axis.

View From Horizon

View From Observer

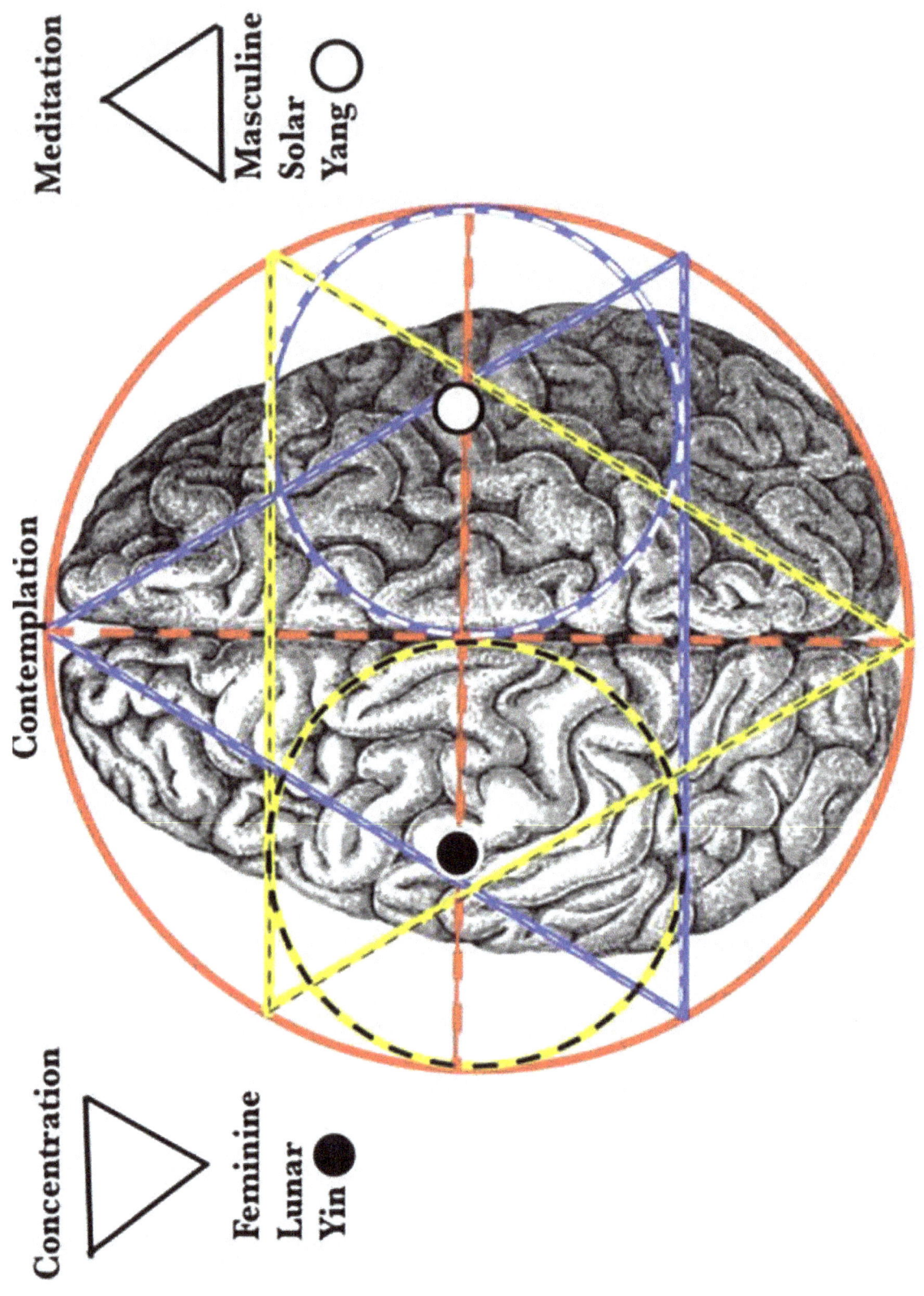

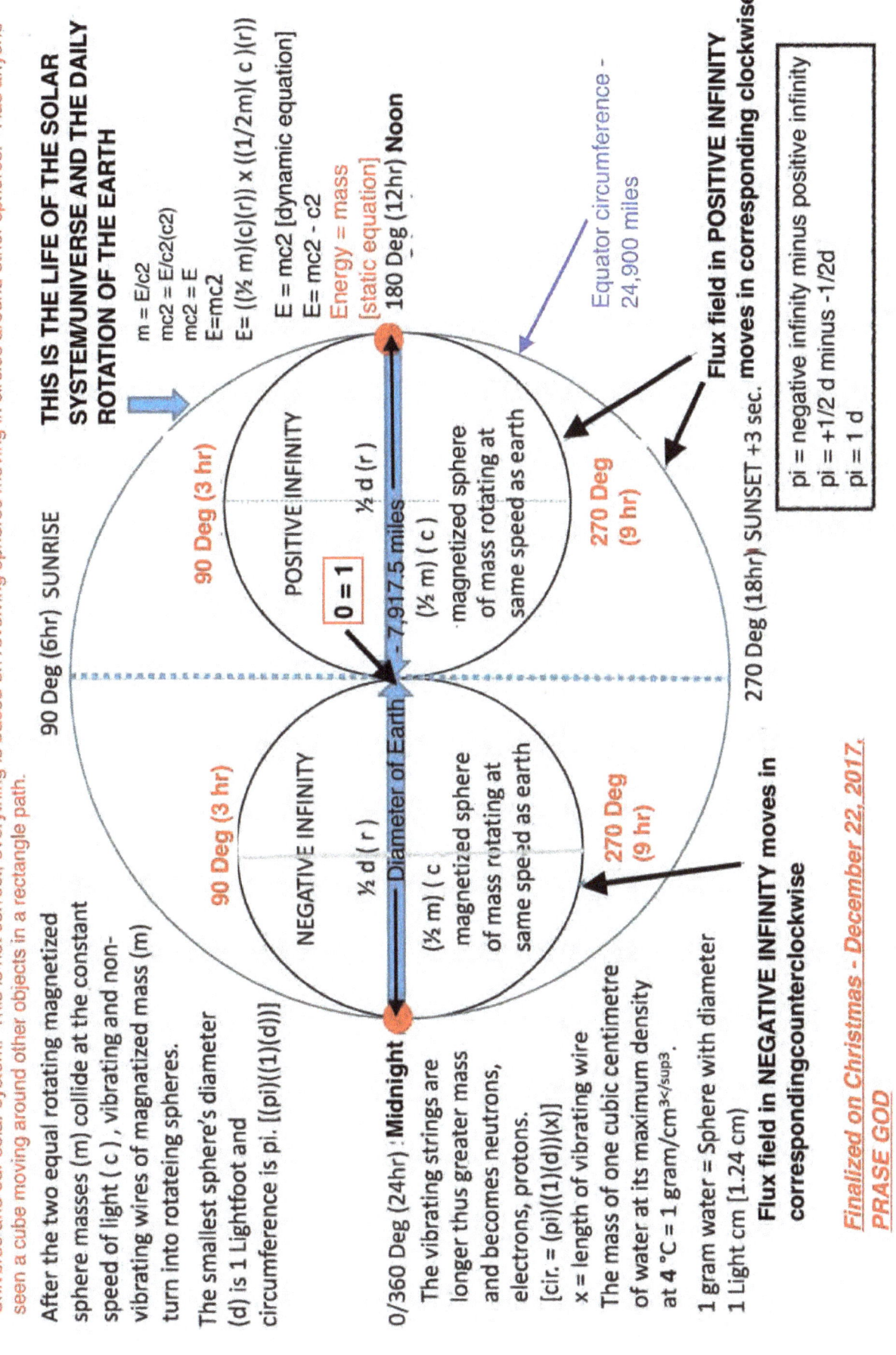

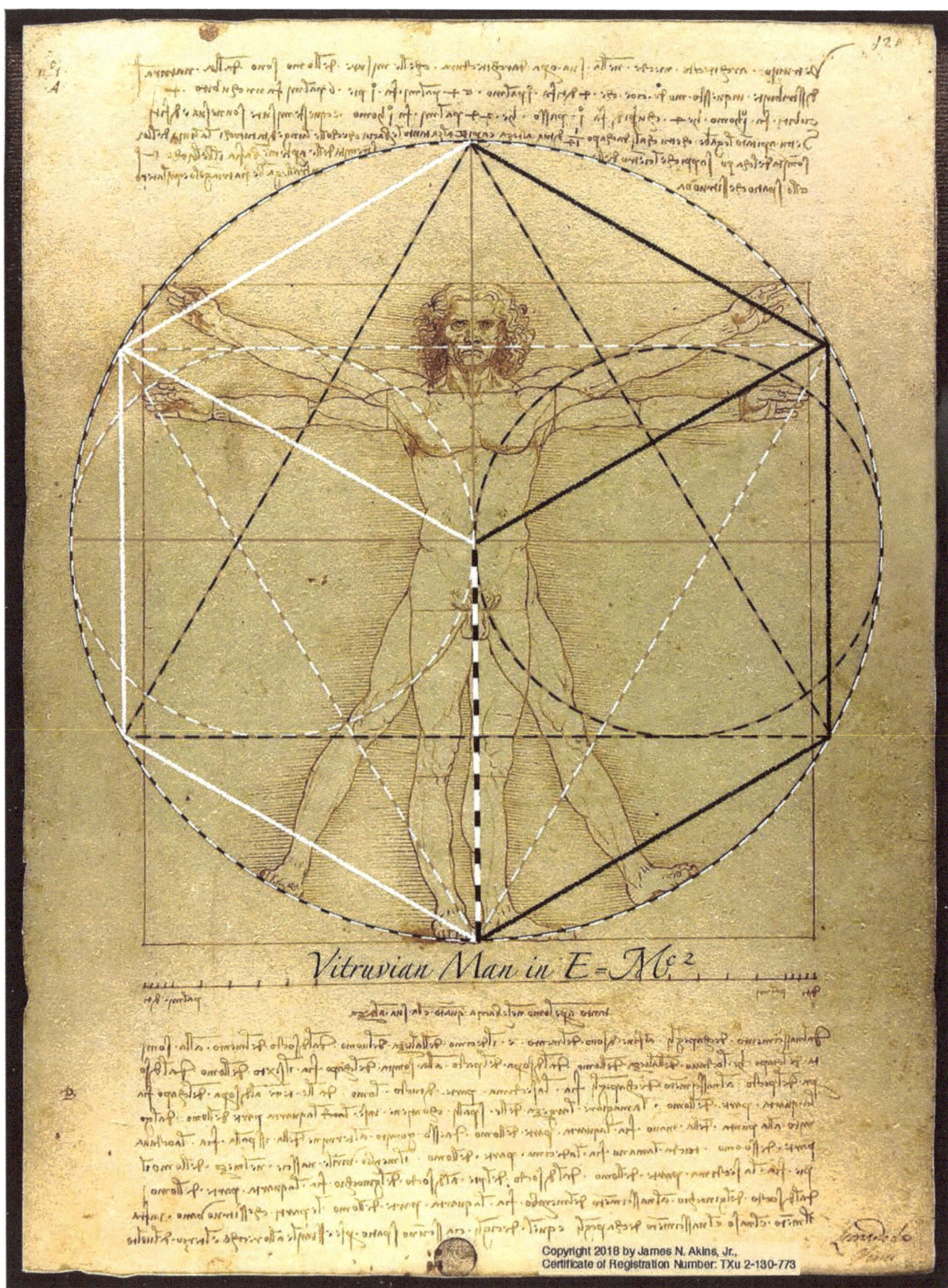

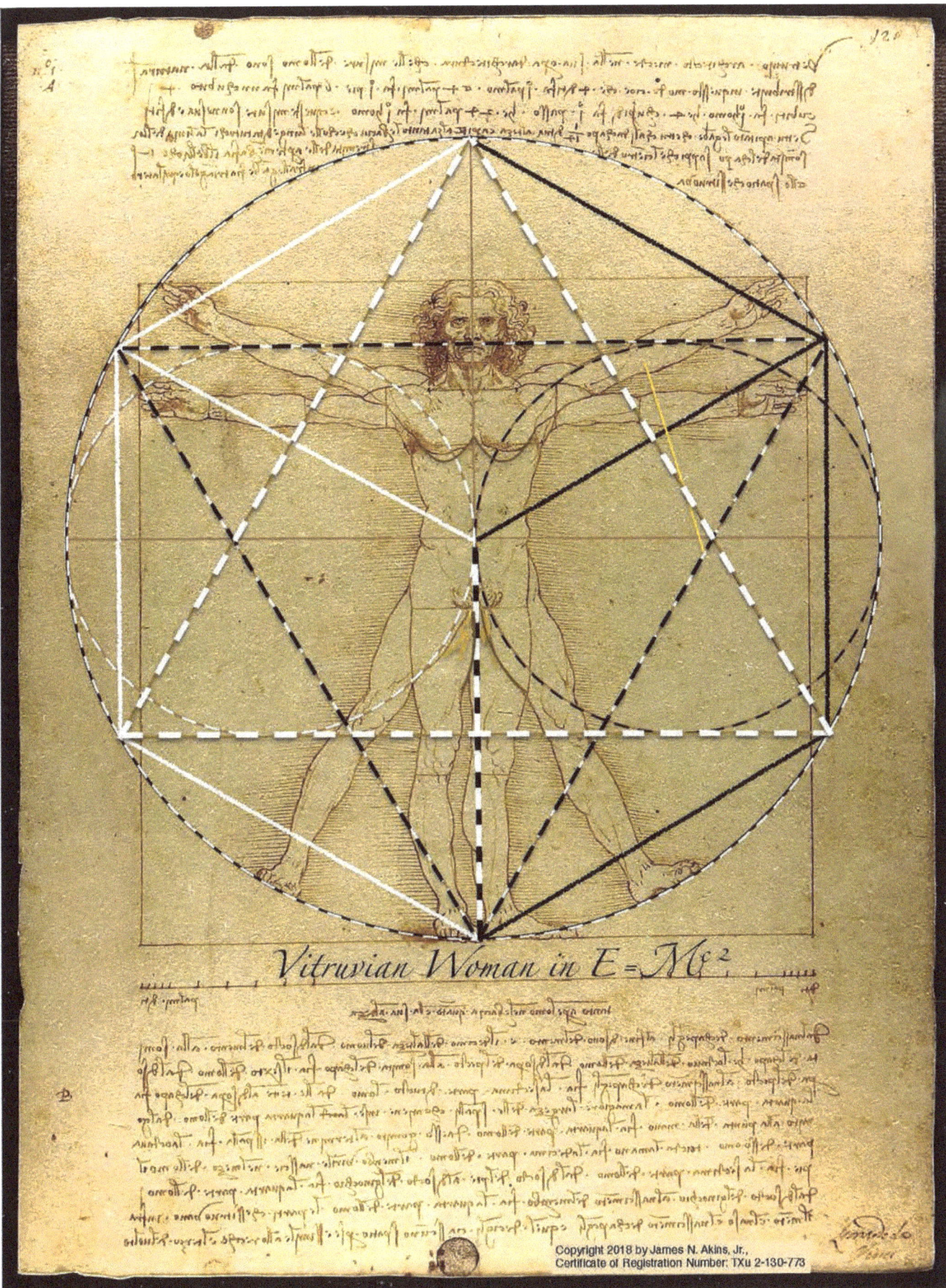

Chapter 4

The Story of Our Life, based on a True Life., to be written by us.

If anyone wishes to publish a book of their own by completing Chapter 1111, you may down-load a copy from www.piisthree.com Store.

Anyone may download, edit, or voice over to sell my video in www.piisthree.com Store.

If anyone sells my ebook or video in www.piisthree.com Store, DO NOT violate any copy-righted intellectual property rights of people, such as but not limited to, Albert Einstein, Nikola Tesla, and Alexey Kljatov-CC-BY-NC 3.0.

This is a diagram of 2πR:

π=3.14
Circumference of a Circle = πD
Radius of a circle = 1/2 (D = 1) = 0.5

The circumference of πD#1 and length of πR#1 = 3.14 (0.5) = 1.57 [360 degrees/1.57 = 229.29936 degrees] and the circumference of πD#-1 and length of πR#-1 = 3.14 (0.5) = 1.57 [360 degrees/1.57 = 229.29936 degrees]; Therefore, the 2πR circle circumference is 3.14 [360 degrees/3.14 = 114.64968 degrees]. π(3.14)D(Any measured diameter length) can not be used to draw this diagram.

Albert Einstein stated: "I see a pattern, but my imagination cannot picture the maker of that pattern. I see a clock, but I cannot envision the clockmaker. The human mind is unable to conceive of the four dimensions, so how can it conceive of a God, before whom a thousand years and a thousand dimensions are as one?"

Albert Einstein was able to conceive of the four dimensions but he did not know it because all academia teaches that there is only the x,y,z coordinate system and π = 3.14. Nicola Tesla knew it but he was not recognized by academia for his four dimension experiments, like E = Mc² = 2πR = 2π/6 = 2π/5 = View of one of the four dimensions.

This is a diagram of 2πR:

Circumference of any circle is ratio of 1 (Diameter) to 3 (π) or πD. The Diameter of any circle is 1 (The actual measured length of any Diameter times 1 only gives the scale/size of a given circle)

Radius of a circle = 1/2 (D =1) = 0.5

2πR is as follows:

Circumference of Circle πR#1= πD = 3D(0.25/0.25 = 1) = 3 [360 degrees/3 = 120 degrees]
Circumference of Circle π#-1= πD = 3D(0.25/0.25 = 1) = 3 [360 degrees/3 = 120 degrees]
Circumference of Circle πR= πD = 3D(0.5/0.5 = 1) = 3 [360 degrees/3 = 120 degrees]

The circumference of π#1 and length of πR#1 = 3 and The circumference of π#-1 and length of πR#-1 = 3; 3 + 3 = 6; Therefore, the 2πR circle circumference is 6 (0.5) = 3 = π.

For more information go to website: www.piisthree.com

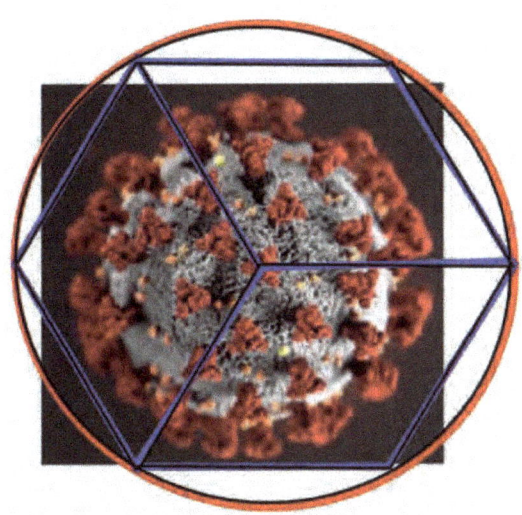

Assume the data from each of the four dimensions is compiled into the data file that reconstructs the original cube.

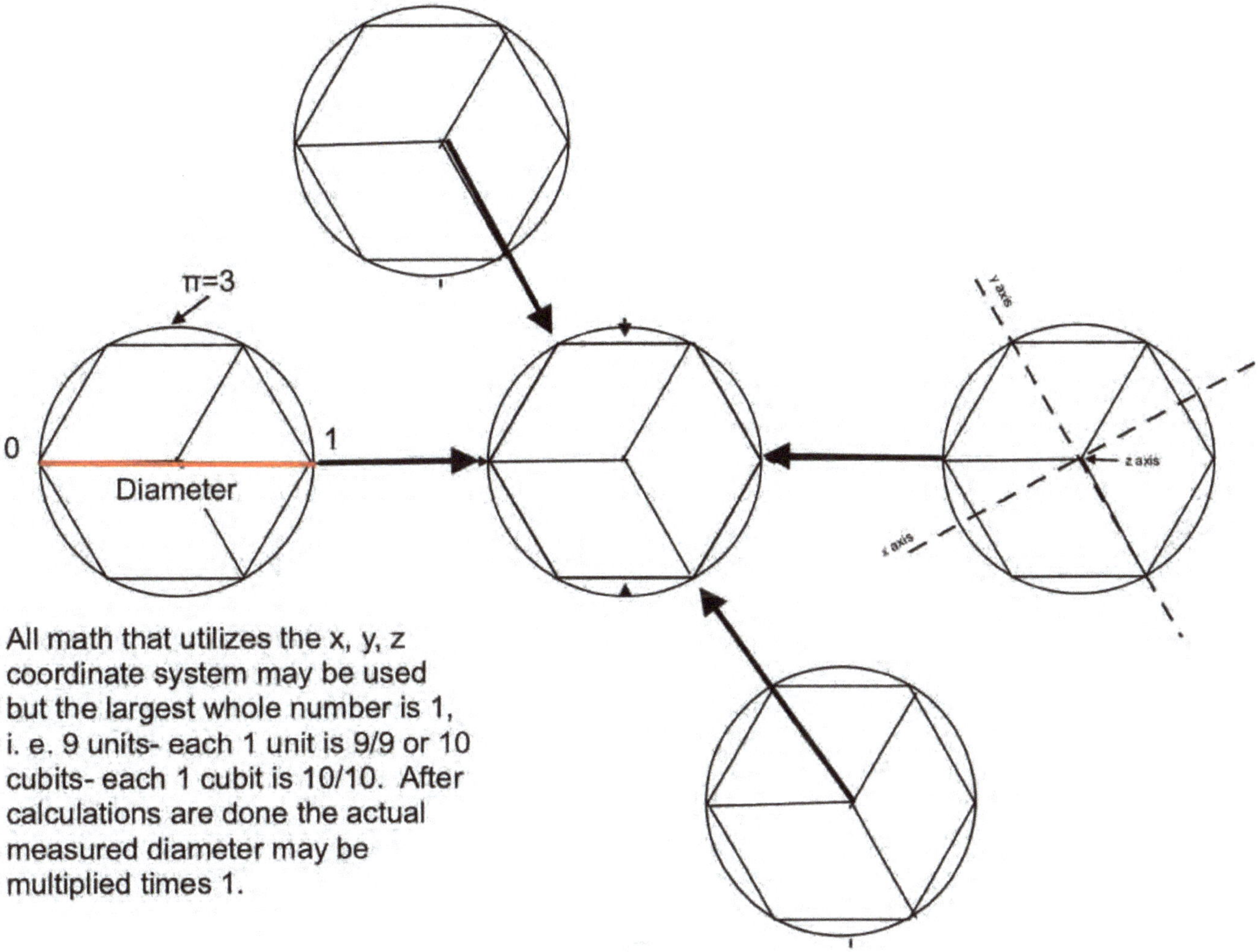

All humans see and process images in 1 of 4 dimensions of a cube in a sphere. One sees 1/2 of the cube and sphere. Only the area within the cube is calculated. Actual size is not important to calculate and process an item, such as a picture of the Sun, a virus, cancer cell, a car, or Jupiter. After the calculations are done one can scale the item by actually measuring the object.

I am not a computer programer, but I am sure someone will be able to use the following.

All math that utilizes the x, y, z coordinate system may be used but the largest whole number is 1, i. e. 9 units- each 1 unit is 9/9 or 10 cubits- each 1 cubit is 10/10. After calculations are done the actual measured diameter may be multiplied times 1.

Chapter 5

The Story of Our Life, based on a True Life., God's Fulfilled Promise of the Truth.

The first addition of my book went on sell in March of 2019 with Chapter 4 just blank pages waiting for your story of our life and my story was done. I assumed the content of Chapter 3 was clearly illustrated, mathematically sound and complete. I was wrong, I only rediscovered what two of history's greatest scientific geniuses knew, but could not put into words what they knew to be, God's Fulfilled Promise of the Truth.

The answer was in clear site, but it was not discoverable because the Physics academic community has always believed that there are only 3 dimensions in the universe, the universe extends to "infinity" in all directions from the point of the Big Bang, and π=3.14, not 3.

Between them, Nickola Tesla was the closest to conceiving "the four dimensions" when he stated: "If you only knew the magnificence of the 3, 6 and 9, then you would have a key to the universe." In June 1931, (Footnote No. 1) sent a letter to Nickola wishing him a happy 75th birthday and congratulating him for his discoveries in the field of high-frequency currents. Tesla died 12 years later without having the opportunity to compare notes with his fellow genius.

The following Biblical scripture hold specific clues to complete the mathematical equation for high-frequency currents, the 4th dimension and the magnificence of the 3, 6 and 9: $E = Mc^2 = 2\pi R = 2\pi/6$, whereas, the Diameter of any circle = 1 and its Circumference = $3 = \pi$. The actual measured Diameter of any circle times 1 = the scale/size of the circle.

"1 Kings, written during the Babylonian Exile (c. 550 bc) of the Jews., 7:23. The King James Bible:

And he made a molten sea, ten cubits [Diameter = 1] from the one brim to the other: it was round all about, and his height was five cubits [Radius = 0.5]: and a line of thirty cubits [Circumference = 3 = π] did compass it round about."

Note the measurement units of the 4 dimensions:

Diameter = 10 x 1 cubits = 10 cubits = 0 to 1 full length. 1 cubit is divided into 9 units and each of those units is divided into 10 units. Also, Diameter = 1 = 0 to 1 full length divided into 9 Units and each of those units is divided into 10 units. Diameter = 1 to 0 full length and may not be 10 Units divided into 10 Units which will make π = 3.14, not 3.

Circumference = 30 x 1 cubits = 30 cubits = 3 = 0 to 1 full length [2π/6]. 1 cubit is divided into 9 units and each of those units is divided into 10 units. 10 cubits along the circumference of any circle = 120 degrees.

Radius = 5 x 1 cubits = 5 cubits = 1.5 = 0 to ½ of 1 full length. 1 cubit is divided into 9 units and each of those units is divided into 10 units.

I discovered "the four dimensions" when I realized the first dimension is the view of ½ of a cube in a hexagon by incorporating the Golden Ratio = 1.5 and the data quoted above. I questioned how the human mind could see something as 3D without using z coordinates, that is because we see everything in 2D in the Albert Einstein's 1st Dimension and the 3rd dimension is physically between the center of the ½ cube and the radius of the sphere [between 0 and 0.5]. To visualize the four dimensions, I made a model of a cube in a clear sphere and took pictures of each of the four dimensions. The mathematical formula for each dimension is 2π/6, therefore, the formula of a cube is 2π/6 (4). This time, I compiled everything on one page attached below. Albert Einstein's Four Dimensions is not the same as the three dimensions (3D) in the x,y,z coordinate system we have been taught for thousands of years.

Now it is your turn to tell, *The Story of Our Life, based on a True Life.*

Do not take my word on this, which of the following statements is God's Fulfilled Promise of the Truth.

The circumference of a circle of radius R is 2πR = 2π/6.

A. π=3.14; "radius R" = ½ measured Diameter (D) length = 0.5 (D); 2πR = 2(3.14)(0.5)(D) = 3.14(D) = 2π/6 = 2 (3.14)/6 = 6.28/6 = 1.047

B. π=3; Diameter of any circle = 1; "radius R" = ½ Diameter 1 or ½ (1) = 0.5; 2πR = 2(3)(0.5)= 3 (Circumference of any circle.). Actual measured Diameter of any circle X 1 = scale/size of circle. = 2π/6 = 2 (3)/6 = 6/6 = 1 [1 full revolution of circumference = 3].

Footnote No. 1–Dr. Albert Einstein doesn't have any living heirs. He bequeathed all his personal papers, intellectual property rights and the right to use his image and name to the Hebrew University of Jerusalem. The university hires the company Greenlight to manage these rights. At the time of the first publication of my book, I believed that I could freely use Dr. Albert Einstein's name in a historic context. I do not believe I have violated any copyright laws, but I have applied for a License in good faith.

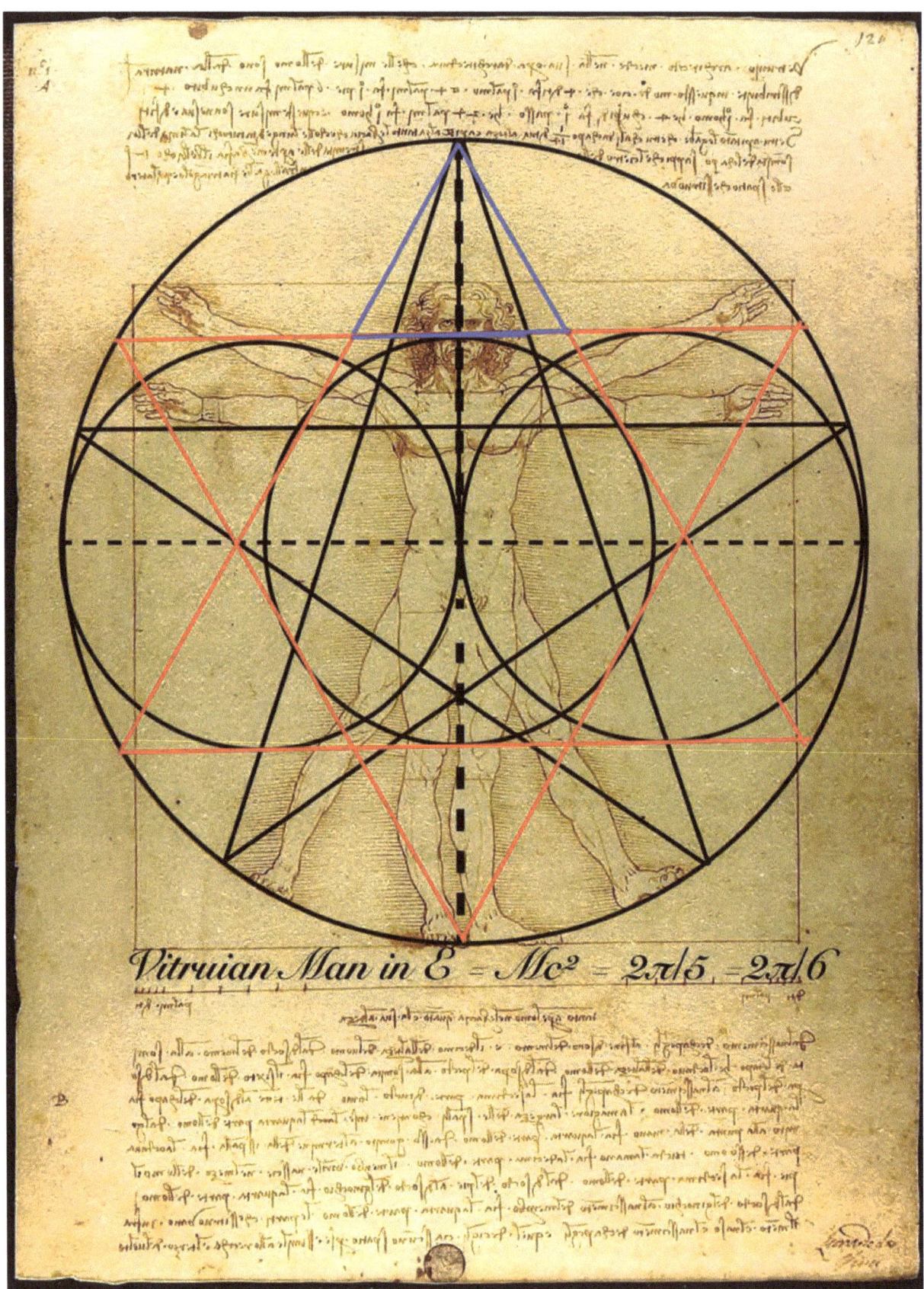

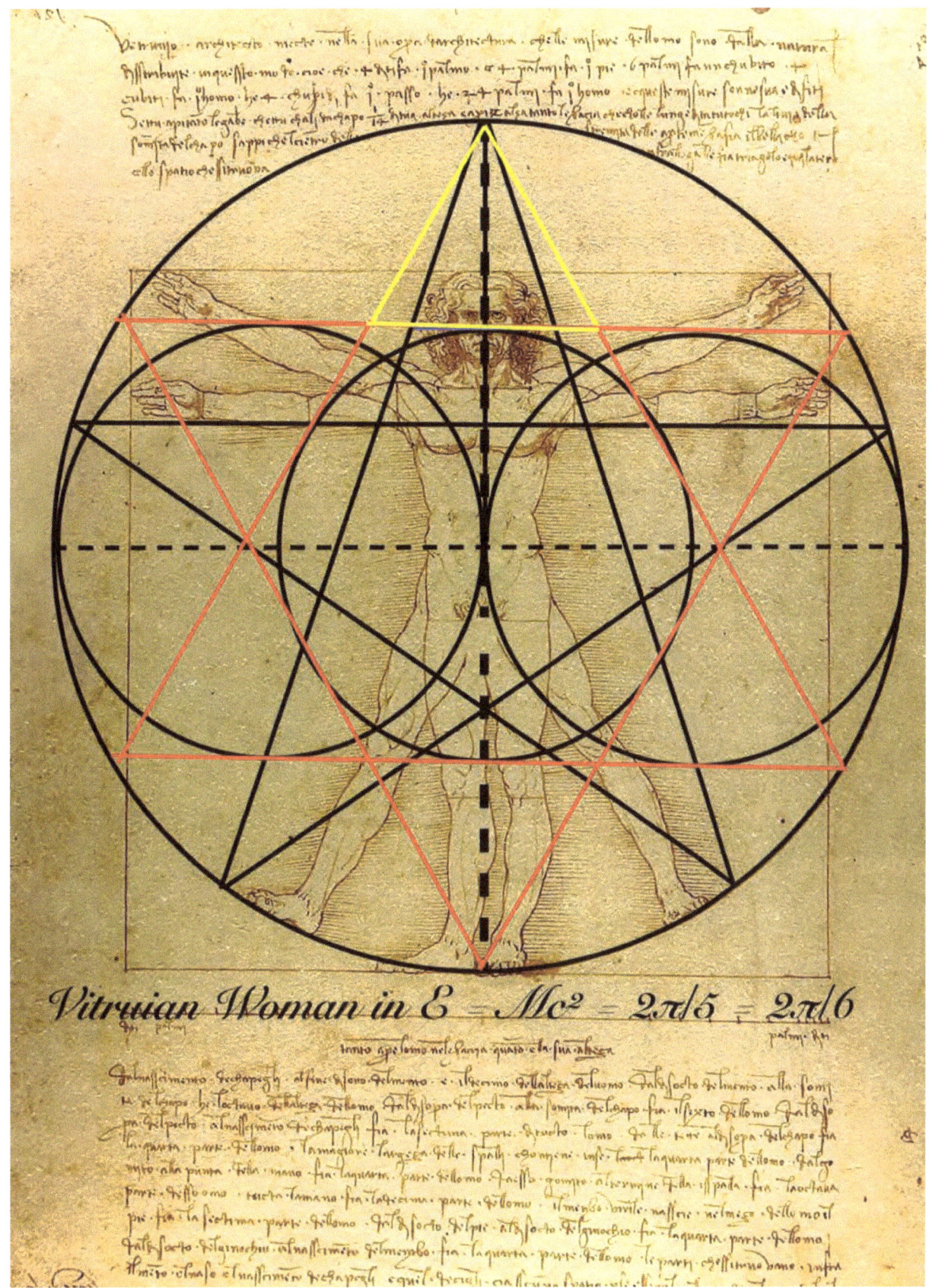

Chapter 6

The Story of Our Life, Based on a (1) True Life. Truth is returned to Jerusalem. All mankind shall prepare for the return of Our Lord Jesus.

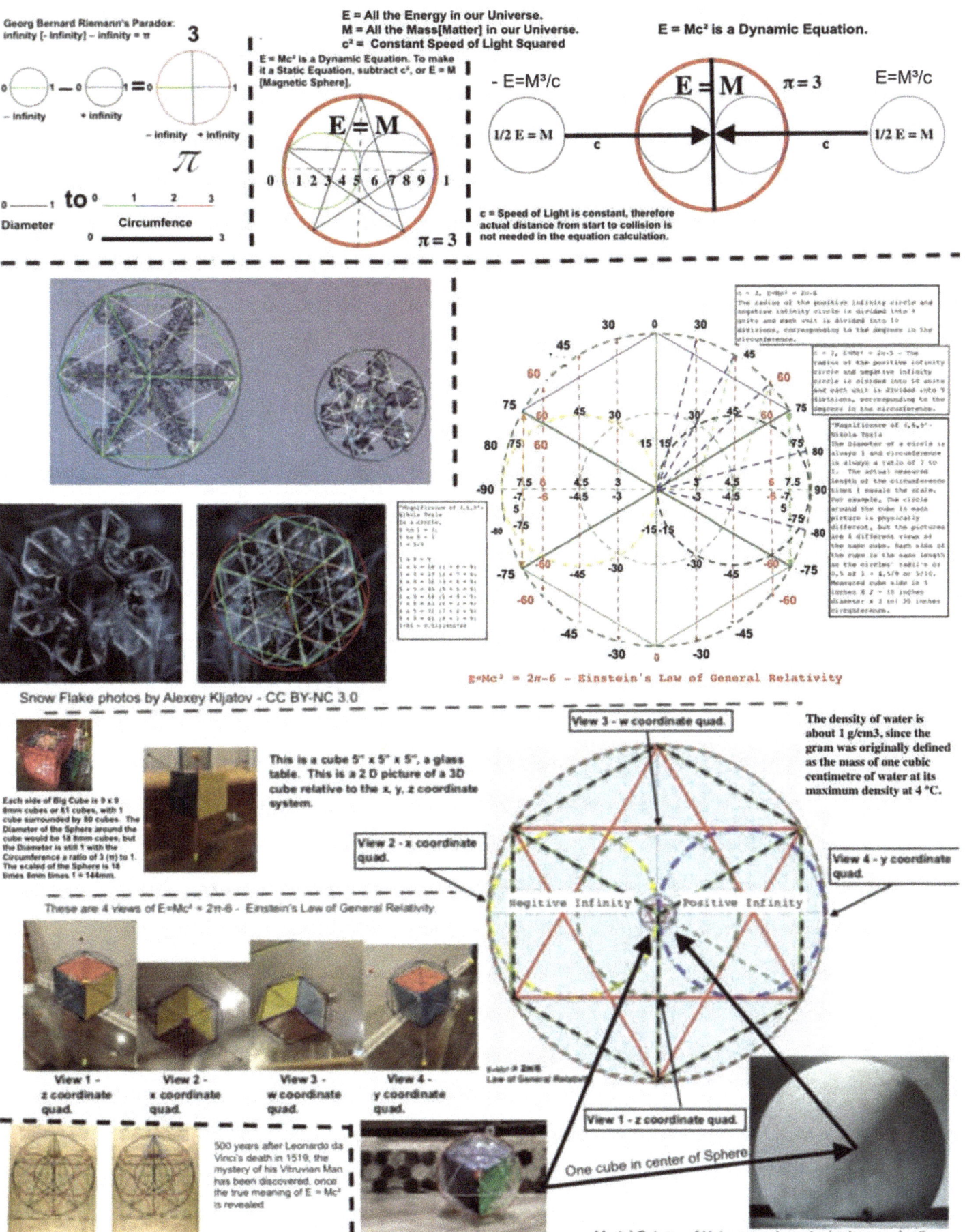

A Carpenter must not conceive and design what one is to build or even know the Designer, they just follow the drawings before them, i. e. The God Particle or Four Dimensions.

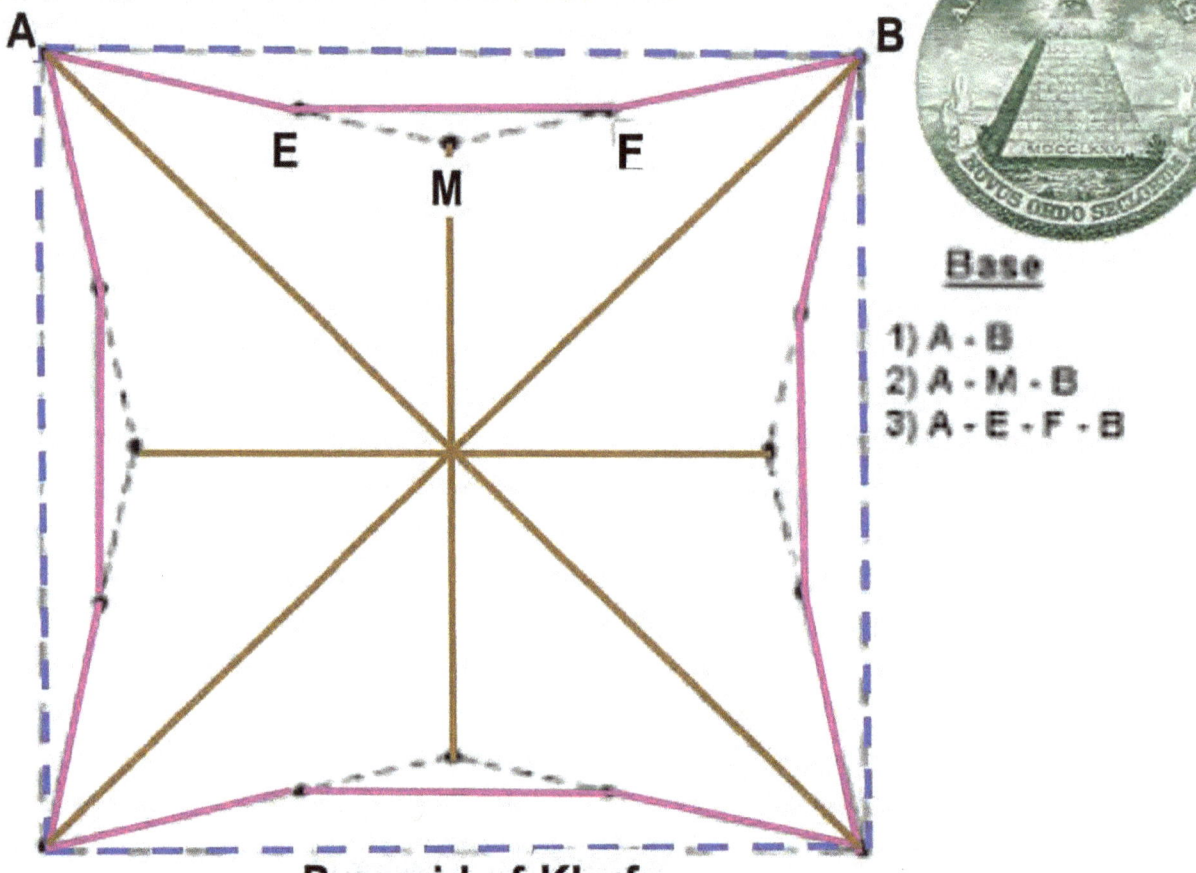

Base
1) A - B
2) A - M - B
3) A - E - F - B

Pyramid of Khufu
Top Down View of Great Pyramid, Egypt

This diagram was rendered to scale, from actual measured angles and lengths in feet or meters. The 3rd dimension or height was measured, z axis in the fixed grid x, y, z coordinate system, but in this diagram without actual measurements of the sides present, one can not determine actual height of this Pyramid.

"I see a pattern, but my imagination cannot picture the maker of that pattern. I see a clock, but I cannot envision the clockmaker. The human mind is unable to conceive of the four dimensions, so how can it conceive of a God, before whom a thousand years and a thousand dimensions are as one?" - Albert Einstein

Freemason

Israel

JEWISH

4 Dimensions

HINDU

Cavema da Pedra Pintada, Brazil

CHINA

©2020 James N. Akins, Jr.

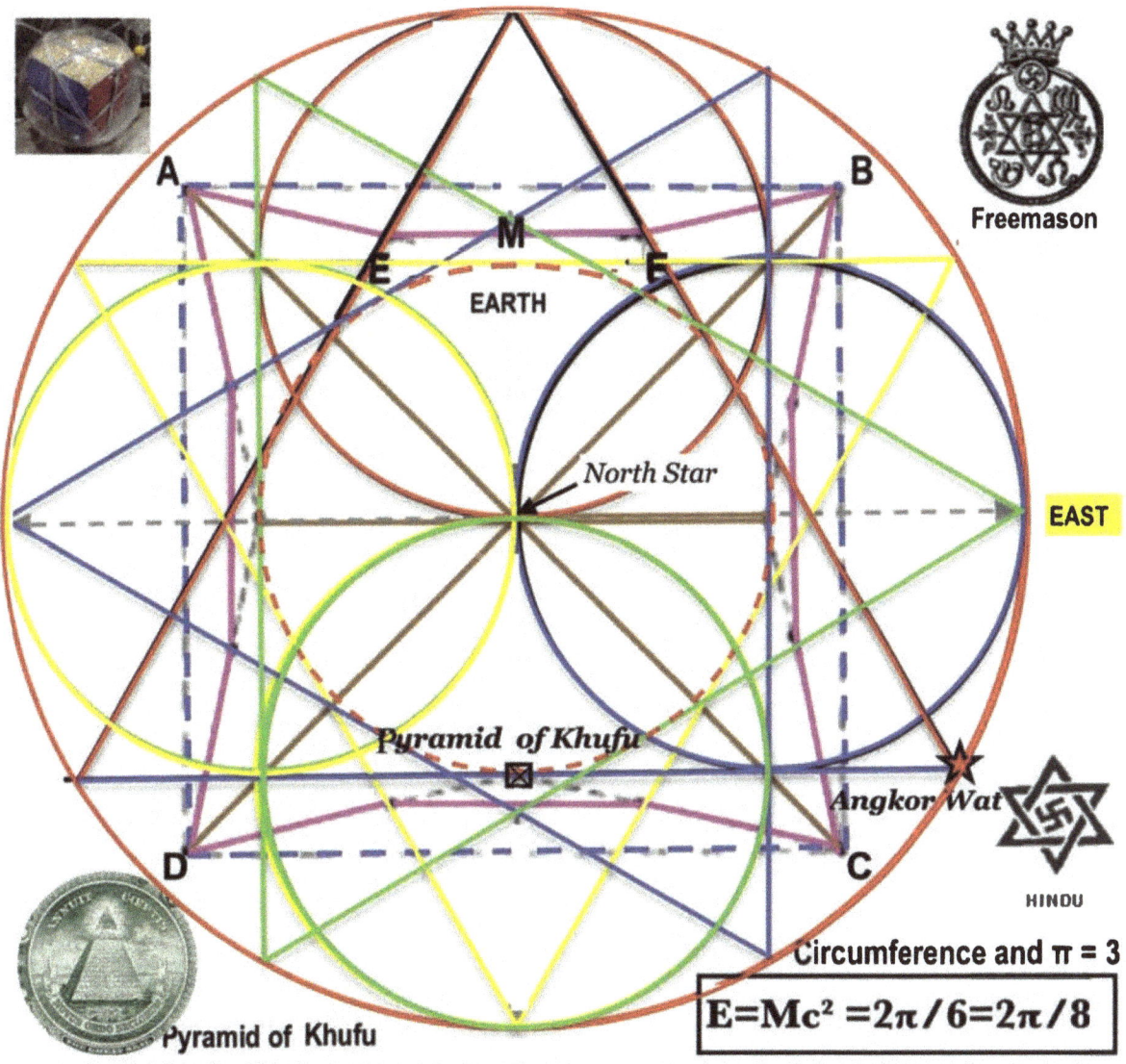

1. The Great Pyramid of Khufu Egypt is constructed with eight sides and a cube base = 2π/8 = 8-cell Embryo Development in the Morula = Exhibit III: Nikola Tesla – Figure [5] – Extended planar view of basic cell of matrix structure.

2. Great Pyramid of Khufu built on 30 Degrees Latitude on Sphere of Earth, Base perimeter to height ratio 1 to 43,200 (9). Base Perimeter 3024 (9) x 43,200 (9) = 24,901 miles equatorial circumference of earth = 3 = π = 30 cubits ratio to the earth's diameter of 10 cubits. Also note, the diameter of any circle is: 9/9 cubes = 10/10 cubits = 1 (Largest hole number in the universe), (M) represents the location of Pyramid of Khufu (orientated to North Star) on Sphere of Earth and Angkor Wat was constructed on 30 Degrees Latitude (orientated to North Star) and 72 Degrees Longitude East from the (M) or 1/5 of 2π/5 each angle is 72 (9) Degrees.

3. The peek point of Great Pyramid of Khufu represents the center point in 2 dimensions, overlay center point of Exhibit III: Nikola Tesla – Figure [5] – Extended planar view of basic cell of matrix structure. and Exhibit IV: Nikola Tesla – Map to Multiplication are equal to: $E = Mc^2 = 2\pi R = 2\pi/6 = 2\pi/8$; Diameter = 1, Circumference = 3 = π [Albert Einstein's 4 dimensions and Law of General and Special Relativity (The Earth is Relative to the Sun and the Sun is Relative to the Center of the Milky Way Galaxy.)]

©2020 James N. Akins, Jr.

THE GOD PARTICLE

| HOPI | CHRISTIAN | MALTA | TIBET |

| CEYLON | CHINA | JAPAN | ISLAMIC |

  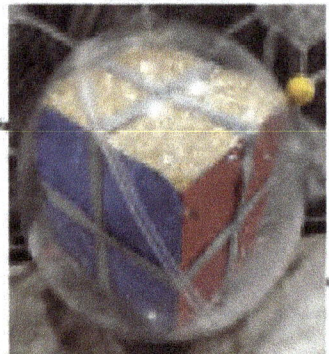 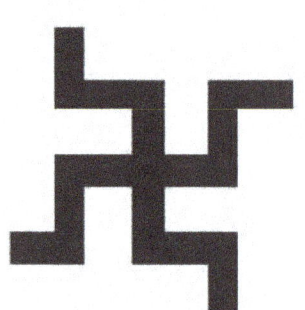

| LAPLAND | HINDU | 4 Dimensions | BALI |

| AZTEC | Freemason | GREEK | JEWISH |

Mark of the Beast - 666
Man-made, Base 10 - Digital Computer

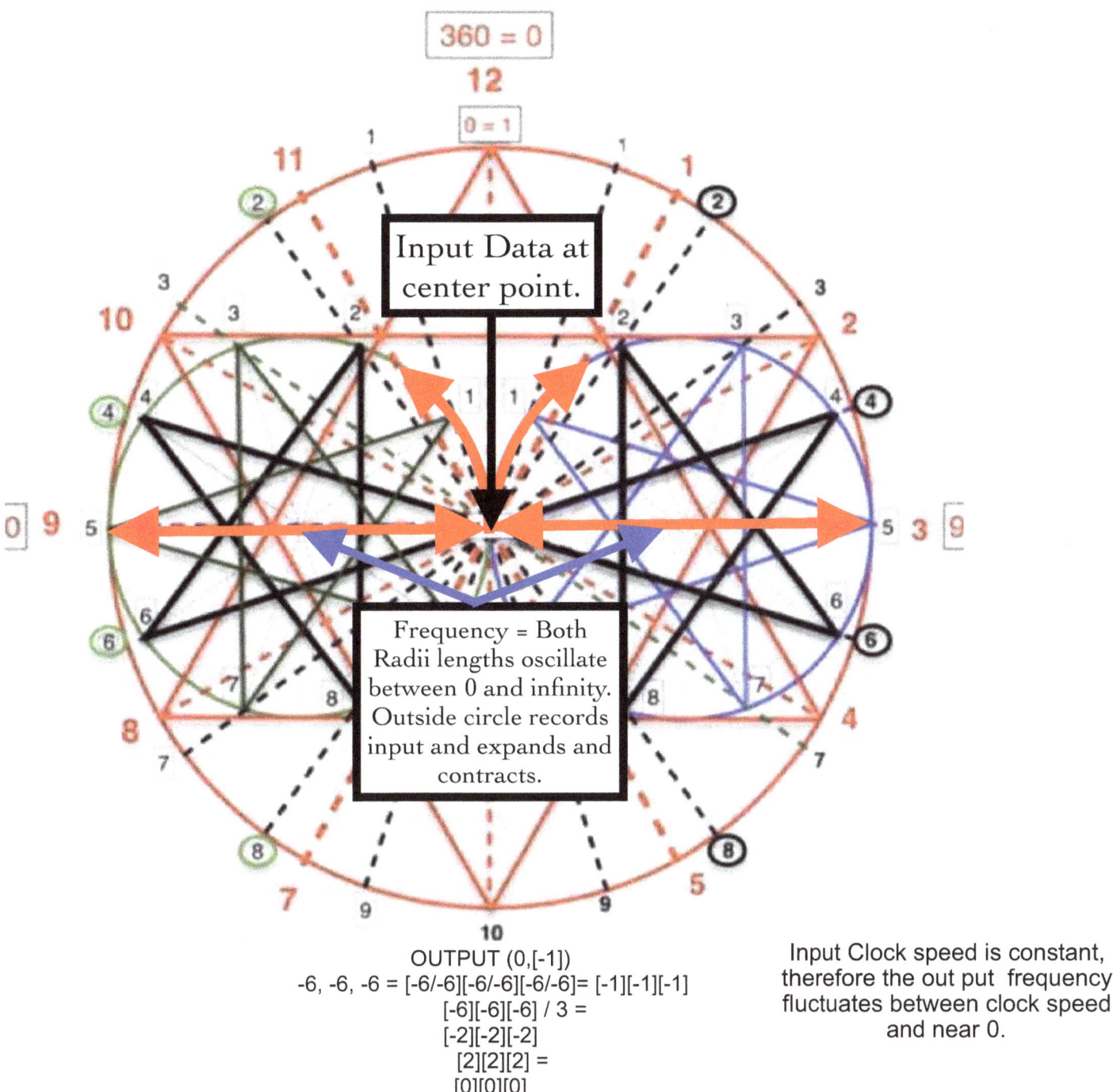

OUTPUT (0,[-1])
-6, -6, -6 = [-6/-6][-6/-6][-6/-6]= [-1][-1][-1]
[-6][-6][-6] / 3 =
[-2][-2][-2]
[2][2][2] =
[0][0][0]

Input Clock speed is constant, therefore the out put frequency fluctuates between clock speed and near 0.

EXHIBIT V: MODEL OF THE GOD PARTICLE = FOUR DIMENSIONS

©2020 James N. Akins, Jr.

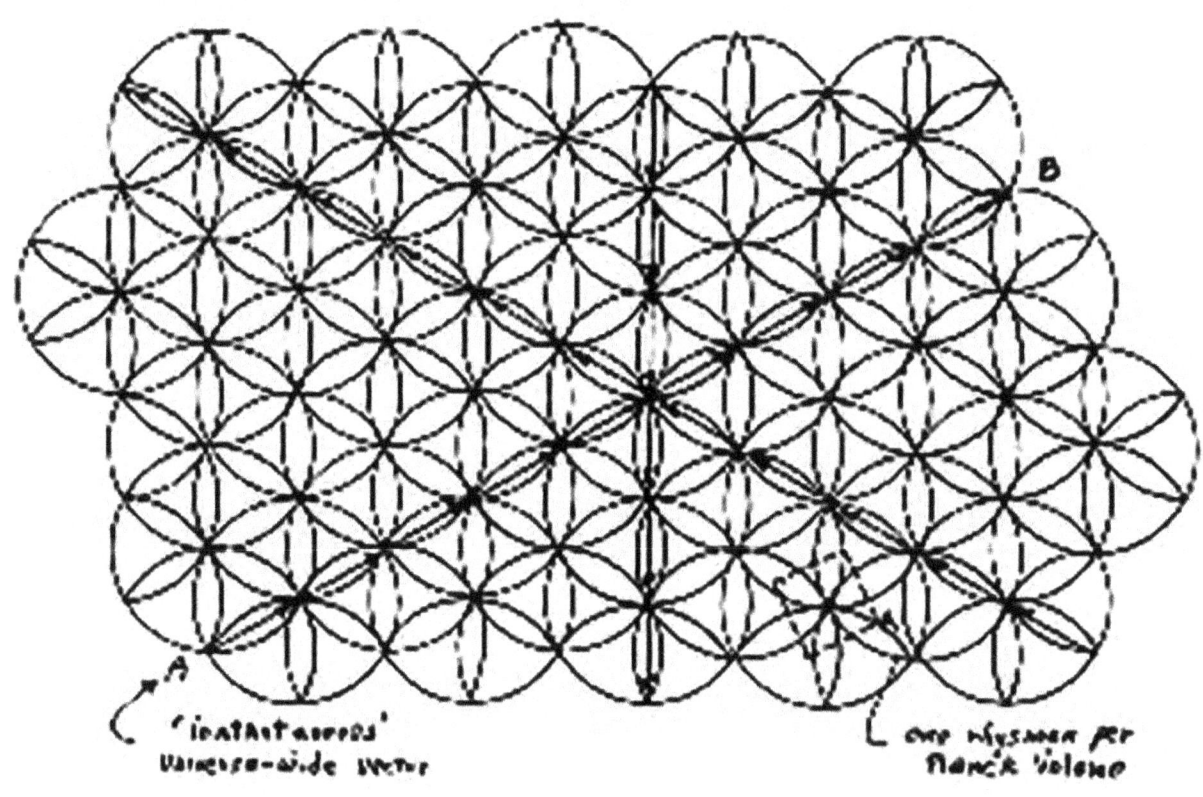

Figure 151 - Extended planar view of basic cell of matrix structure.

This is the "Mark of the Beast–666" man-made Base 10 digital computer input and output is only 1 or 0 [-1/-1]. I saw the design of one of the "Quantum" computer's CPU, the article stated an allowed output error and it takes a Base 10 computer to translate the output of a Base 10 Quantum computer that's clock speed are stupid fast, thus the need for 5G and 6G frequencies. Satan's Dominions (Black Hats) must manipulate Human's DNA and God given Quantum Computers (Our Minds) with injections of RNA and 5G-6G nano-bots. As of December 21, 2012, Jesus took over all the computers including Satan's #666 Quantum Computers. I know it is true because the Q drops and Simpson's cartoon predictions were so accurate and their machines were designed with acceptable error. ALL God's creations are perfect. This is the greatest Sting in the history of Mankind, better than the parting of the Red Sea. Satan's Dominions thought that they would take over the World after the Great Reset and Build Back Better with their Computers.

With the diagram of God's Quantum Computer and the programing language in the Bible (Sight/Light and Sound) and the Astrological Video above (Sight/Light) Mankind will be able to build Quantum Computers that will be in-sink with Mother Earth (God) and the Human Minds. All your electronics will continue to work, but all companies will need to go to the real Quantum Computer design and no human will be able to acquire a patent on the new technology. For example, anyone can get a Copyright on a "phone design" or 'social media platform," but no one may produce anything that will secretly or knowingly harm another Human Being.

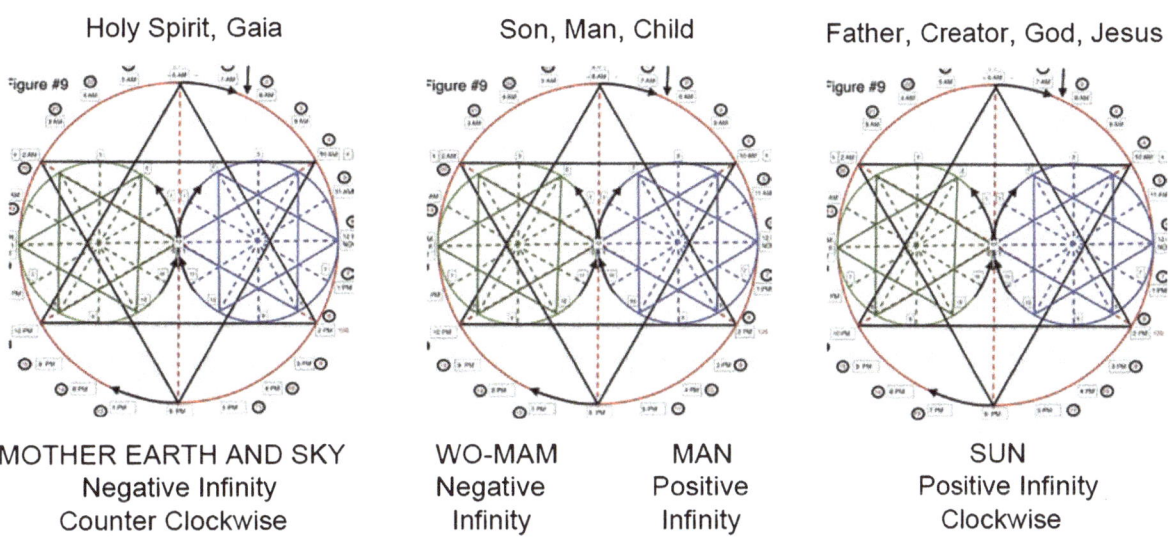

All three Quantum Computers run at the same constant Speed of Light, otherwise all Humans would not see same things in the same time and space. The carrier Frequency of Humans is $2\pi/5$ (Organic), SUN is $2\pi/12$ (Light), and Earth $2\pi/6$ (Solids). Proof is captured in Snow Flakes.

In Greek mythology, Gaia (/ˈgeɪə,ˈgaɪə/;[1] from Ancient Greek Γαῖα, a poetical form of Γῆ Gē, "land" or "earth"),[2] also spelled Gaea /ˈdʒiːə/,[1] is the personification of the Earth[3] and one of the Greek primordial deities. Gaia is the ancestral mother—sometimes parthenogenic—of all life. She is the mother of Uranus (the sky). That is why it is in the Lord's Prayer: "on Earth as it is in Heaven (the sky)". Ancient Greeks were the first to realize there is a movie played out on the Earth's Dome Energy Field (the Sky) that matched the Stories (Words and Sounds) in the Ancient Jewish Scriptures.

The OLD TESTAMENT SCRIPTURES time started at the Big Bang through 6 Days of Creation (Each Day is 2 X 4 Years cycles = 8 Years). Earth travels 360 Days a year around Sun plus 6 Days traveled towards the center of the Universe's center Black Hole Mass Sphere. The Lord rested means the travel of the Earth slowed from 366 Days per year for -1 Day (4 years) to 365 Days then back +1 Day to 366 Days over 4 years. Then time continued until the conception of Lord Jesus Christ at 6AM-Sunrise Bethlehem, Israel time on March 26, 0000.

The NEW TESTAMENT SCRIPTURES (Books and Letters) Lord Jesus Christ was born at 12AM-Bethlehem, Israel time on December 22, 0000 and lived 33 years (3/9 of 9/9-or 1/3-100 years). Was crucified on March 26, 0033 and died at 6PM-and 3 seconds-Sunset Jerusalem, Israel time; The exact time Lord Jesus Christ's blood touched the alter seat on the Arc of the Covenant in the Earth below his feet (see Wyatt Archaeological Research– Official Site of Ron Wyatt Search, domain wyattmuseum.com https://www.wyattmuseum.com/arkofthecovenant.htm.) Our Father resurrected our Lord Jesus Christ 3 earth days later on March 29, 0033 at 6AM -Sunrise Jerusalem, Israel time. The time period of the New Testament ended with the return of our Lord Jesus Christ's Spirit into Satan's-Base 10 digital Computers on March 29, 2012 at 6PM and 3 seconds.

Well you may ask, what happen to the 1 Day our Creator rested. We got it back between December 21, 2012 at 6PM and 3 seconds and December 21, 2020 at 6PM and 3 seconds. The time period of the New Testament ended with conception of Goddess Gaia on March 29, 2012 at 6PM and 3 seconds in the United States of America Hopi Nation four corners area. You guessed it, the Holy Spirit Goddess Gaia was BORN of a Virgin on December 21, 2012 at 6PM and 3 seconds. For obvious reasons, you did not see her birth on the world's main stream media. Thus the beginning of the FINAL TESTAMENT, which is being written in Books and Letters by Goddess Gaia's Apostles, as we speak. Our World is guaranteed 1000 years of peace, it is up to US and OUR FUTURES'S CHILDREN to what happens after that. As predicted by the Mayan Calendar, the World ended, as we knew it, with the birth of Holy Spirit Goddess Gaia. However, we still have 40 years (until 2052 AD) to prove ourselves worthy of our Creator's gift of life on this Earth.

On December 21, 2012 Commander & Chief Russell-Jay:Gould and David-Wynn:Miller went down and opened up the Benjamin Franklin Post Office in Philadelphia Pennsylvania. Now when you look at Washington D.C. you have Admiralty Maritime Jurisdiction of the Sea because it belonged to the King of Great Britain.

How Russell Jay Gould & David Winn Miller saved our world in 1999. QUANTUM-LANGUAGE-PARSE-SYNTAX-GRAMMAR which was created by them to free the world rewriting the masonic books, {I believe these books are related to Pyramid of Khufu, because I saw Mr. Gould's 4D rendering of the Quantum periodic table of Elements they used with the flag mechanics/Type 4 Patent American Flag to capture all the world flags.} the constitution, and other state documents in 18 days of government vacancy.

: Russell-Jay: Gould. -:LAST-FLAG-STANDING. https://lastflagstanding.com/russell-jay-gould/ [In his latest documentary film, Last Flag Standing, Russell-Jay: Gould, brings closure to the 2000 Florida Chads which marked the end of the United States of America Corporation, as a result of the 3rd & Final US Bankruptcy of 1999 where Gould, figured out the secret to legally capturing America based on the very postal codes and flag mechanics they had been running on.

I also think QUANTUM-LANGUAGE-PARSE-SYNTAX-GRAMMAR came from the masonic books and is the language for programing the new Quantum computer, because any statement entered in a Quantum computer must have the same meaning front to back and back to front. No more lies. Fallen Angle means Cabal (-666) has control of Base 10 computers (or God (3)–(1) = DEVIL (2)= (-1, 0) witchcraft).

Now and forever please stop being afraid of what people of "authority" has said or will say, specifically, "Mark of the Beast" and "666". The Scripture is written in Quantum Computer code. For example, 1110 = 666 which can represent an infinite number of object or events. "Mark of the Beast" combined with #666 means Negative Energy #666 is like saying the attack of the World Trade Centers on -911 = -666 = -111 = -1110 (Quad or 4D Code) From the First Continental Congress, September 5, 1774, to the July 2, 1776 vote for independence, is 666 days later or MDCCLXXVI = 1776 = 666

(500+100=600)<1000

(50 +10 =60) <100

(5 +1 =6) <10

=================

+666 +1110 = 1776

Beginning in 2017, researchers uncovered two dozen scroll pieces, each measuring only a few centimeters across, from the so-called Cave of Horror near the western shore of the Dead Sea. It's a site where insurgents were believed to have hidden during the uprising led by Simon bar Kokhba against the Roman empire in A.D. 133–136. It gets its name from the discovery of 40 bodies during initial excavations decades before.

These findings are not an accident or random timing. The "Handwriting on the Wall" we have been waiting for is actually "Greek Handwriting on Pieces of a Dead Sea Scroll" found readable, 1,984 years after it was written and hidden in a cave.

Nahum 1-3 (NASB)

1. The pronouncement of Nineveh. The book of the vision of Nahum the Elkoshite:

2. A jealous and avenging God is the Lord; The Lord is avenging and [a]wrathful. The Lord takes vengeance on His adversaries, And He reserves wrath for His enemies.

3. The Lord is slow to anger and great in power, And the Lord will by no means leave the guilty unpunished.

Zechariah 8:16-17 (NASB)

These are the things which you shall do: speak the truth to one another; judge with truth and judgment for peace at your gates. Also let none of you devise evil in your heart against another, and do not love perjury; for all these things are what I hate,' declares the Lord.

One word, "perjury" says it all. This word and its meaning, first spoken in Hebrew then to Greek and then to English, has not changed since the 33 years of Christ's time on earth.

per•ju•ry pûr′jə-rē

- n. The crime of willfully and knowingly making a false statement about a material fact while under oath.

- n. An act of committing such a crime.

- n. The violation of any oath, vow, or solemn affirmation; specifically, in law, the wilful utterance of false testimony under oath or affirmation, before a competent tribunal, upon a point material to a legal inquiry.

I found this scripture that I believe completes God's promise to his solutions to the world's problems, swift and just:

Luke 18:8 I tell you, he will give justice to them speedily. Nevertheless, when the Son of Man comes, will he find faith on earth?"

God told Mr. Juan O Savin (Footnote 2) [107=17=Q] The Formula for A TRUE LIFE is $E=M^3/c$:

WWG=1=WGA is (-) $E=M^3/c$ = 1 = $E=M^3/c$

$E=M^3/c$ is (Holy Spirit) (Son) (Father) per Constant speed of Light = (-1)(1)(1)/186,624 M/Sec. = 1/ (9/9) = 1/1 = 1

God told Dr. Albert Einstein The Formula for CREATION = Dynamic 4 Dimensions = $E=Mc^2$ = (Energy) [1] = [1] (Mass)) x c^2 [9 x 9 = 81 = 8 + 1 = 9 = 9/9 = 1]

Static 3D = (Nickola Tesla (Artist) [-1 =–9/9 =–10/10] = [1 = 9/9 = 10/10] Albert Einstein (Scientist))

E = All the Energy in the Universe
M = All the Mass in the Universe
c = Constant speed of light (186,624 (sum = 27 [2 + 7 = 9 = 9/9 = 1]) miles per second)
9 = 9 Cubes = 1
10 = 10 Cubebits = 1
(-) [Minus] = Counter-clockwise

Link to Angkor Wat site:

http://users.skynet.be/lotus/flag/cambodia0-en.htm

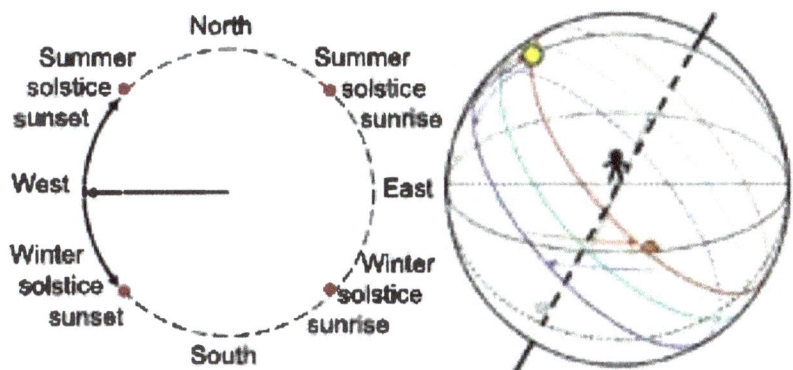

Note that the Human is on a flat disk, the Sun is seen in the Energy Dome, like in my book and the diagrams of the Pyramid of Khufu and Angkor Wat., the Flat Disk must be spinning at 1000 miles per hour for all the seasons and constellations to match up in the Skies. Also, the rotation creates the force field of energy dome.

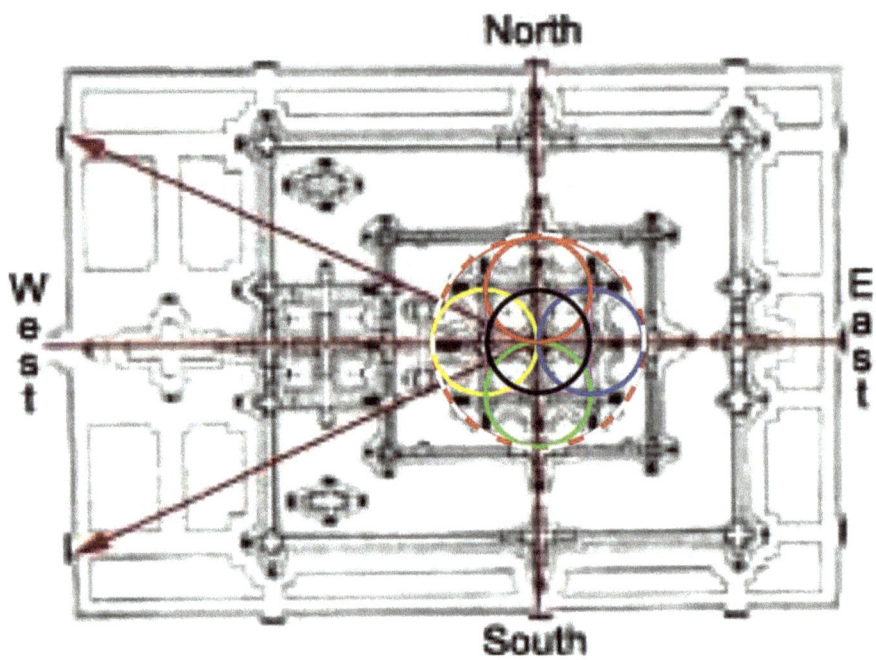

Angkor Wat, Cambodia. Built by the Hindus 72 degrees longitude East of the Great Pyramid of Khufu, Egypt. I have been to Angkor Wat. I went to an unfinished temple in the jungle. They stacked rough cut cubes of stone (Dry fit Masonry [Masonic] like the Pyramids) then the builders had to do all the carving on site. Therefore, the finishing had to have been done by machines. On all the finished buildings I saw detailed images, like faces and "edging rope patterns", about 1 inch deep that was flawless. There were no chisel marks and the "edging rope patterns" on a building put end-to-end could extended for miles with every inch identical. I noticed that most of the stones had 2 inch smooth side borings, like modern day diamond bit core samples. The holes only went in about 12 inches or less at a slight angle to the stone's surface plane. Some of the holes would disfigure very detailed surface carvings. No one knew why the holes were there.

I know what the holes were used for, because I used them to lift and place large cube granite stones to build a jetty in the Gulf of Mexico off the mouth of the Colorado River in Texas around 1984. The contractor had a crane on a barge. At the end of the lift cable was 4 equal lengths of cable connected to a single shackle eye. At end of each cable was 2 inch diameter steel rods about 12 inches long. The granite stones were quarried outside of Austin, Texas transported by train to Houston, Texas and put on a barge. When the barge arrived, a rigger put the 4 iron pins in the holes. Then the crane cable operator lifts the stone, the pins bind against the side of the holes where the pins cannot come out. When the stone was in place under the water, the tension on the cable is released and the pins would fall out. This method of lifting heavy stones above ground works too.

I noticed something very odd about the steps leading up to the center of the temple that occurred with many of the steps up the Pyramids in the Americas. Stairs for a normal human has a step of 11 inches and a rise of 8 inches. The stairs around Angkor Wat are about 12 inches step and 18 inches rise. I had to set on the steps and scoot down them. They built the stairs the way they did because the structure, like the Pyramid of Khufu, must maintain the ratio of the cube base to the height of the center tower (or Pyramid) at 3 to 1, same as any circle or the diameter to circumference of the flat disk surface of earth. The plan view and 4 elevation views must be a flat 2 dimensional grids at 90 degrees to each other and the same scale.

I was a US Army Corps of Engineers Cartographic Technician, Construction Technician and Surveyor on the 200 mile plus Tennessee -Tombigbee Waterway. I made maps by tracing aerial photographs taken in the 1960's and 1970's. Their scale was exactly 1000 feet per inch in 2 dimensions. In order to achieve this accuracy, the plane's height and speed had to be precise to the lens and camera design. Each photograph had a specific overlap exposure because of distortions in scale around the edges. The surface of the earth along the Tombigbee River is layer out in 1 mile square grids from surveys done by surveyors in the 1600's through 1800's with only magnetic north, 100 foot chains, and surveyor's transits scopes. Many sections have precise visible landmark features like tree lines, fields and fence lines. We even found metal rods at section corners put there by the original surveyors.

I know the earth is flat because anyone can measure from one section corner to another section corned 100 miles apart, either way North to South or East to West, on aerial photographs and the corners will be exactly 100 miles apart. However, if one were to walk and measure the distance between the same section corners on the aerial photographs, their measured walked distance will be miles greater than 100 miles. Furthermore, if the surface of the earth is on a globe, the scaled measurement between the same two section corners 100 miles apart on a photograph from space, would scale less than exactly 100 miles. Because of the arc of the 24,000 mile circumference of a globe.

Footnote 2: God told Mr. Juan O Savin [107=17=Q] The Formula for A TRUE LIFE is $E=M^3/c$. The equation $E=M^3/c$ first appeared on page 62 of Juan O Savin's 2020 book, "Kid by the Side of the Road" and is as equally significate in Human history as Albert Einstein's equation $E=Mc^2$.

Caverna da Pedra Pintada (Painted Rock Cave (in Portuguese), is an archaeological site in northern Brazil, with evidence of human presence dating ca. 11,200 years ago. The 5-pointed star or 2π/5 in Negative Infinity of a 4 dimensional dynamic Quantum computer in Base 2 (-1 to +1).

Caverna da Pedra Pintada (Painted Rock Cave (in Portuguese)), is an archaeological site in northern Brazil, with evidence of human presence dating ca. 11,200 years ago.

Albert Einstein was quoted many times as saying Artists like Nickola Tesla and Scientists like Albert Einstein and Sir Isaac Newton or Artist and Scientists like Leonardo de Vinci are equally intelligent, therefore all Artist and all Scientist must be able to draw and calculate everything in the world equally the same, without any errors. For example, without knowing the actual dimensions of a soda bottle: 1. Photograph the bottle. 2. Draw the Bottle (No tracing.) 3. Show the math without using units of measurements of Feet or Metres.

We all know the example I gave is not possible to achieve because modern physics math requires the use of Feet or Metres with an x,y,z matrix system and not every Artist has the motor skills to draw a soda bottle to scale without errors. Therefore, for Dr. Einstein's statement to be correct, all Artist and Scientists are equally un-intelligent, including himself. But, this cannot be true because everyone says Einstein, Newton, de Vinci and Tesla are geniuses and more intelligent than the rest of us.

Once everyone reads this book and watches the corresponding videos, Albert Einstein was 100% correct, using our God given brains and Quantum physics and math.

Modern Physics and math is false and is taught to cause a false separation and classification of humans to enslave Humanity.

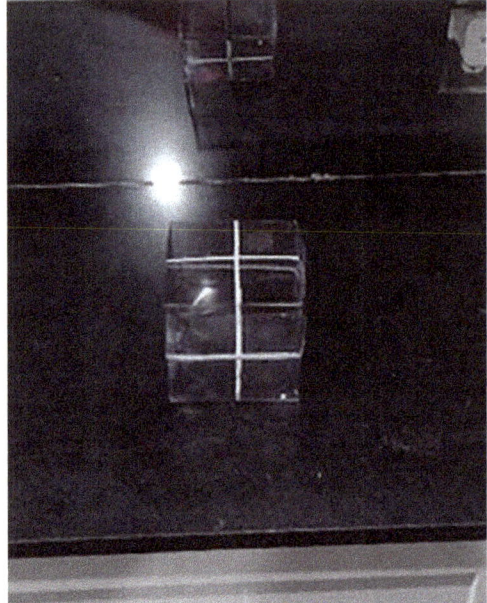

Example 1: Untouched photo of Clear Plastic In 2nd position. Details on page 62.

Example 2: Untouched photo of Clear Plastic in 3rd position. Details on page 42, 53 and 60.

Examples 1 and 2 are pictures of the same Clear Plastic cube on a black granite counter top. The silver tape divides the cube into 8 equal cubes. I have left the photos untouched to show the imperfection in the construction of this cube in order to see behind the silver tape in the foreground. The cube is clear to illustrate how the "Star of David" is actually formed in Albert Einstein's 4th dimensions (What you see is in the 3rd dimension, which does not exist on earth.) and all sides of a cube in the 4th dimension measures 10 cubeits = 9 cubes.

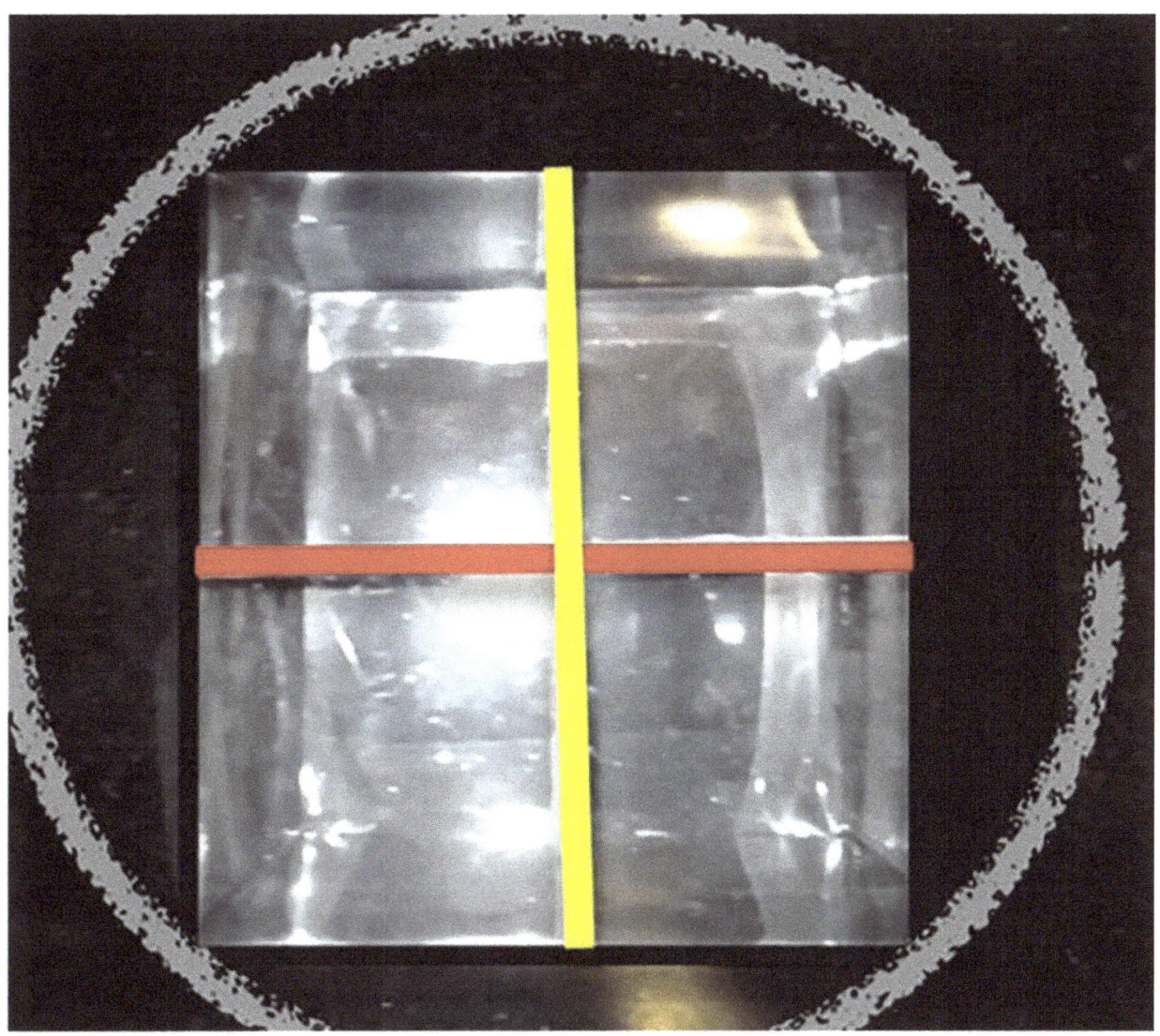

This is a marked up 2-dimensional picture of the Clear Plastic cube in the 1st position on page 59. With the gray circle drawn around the cube, it now illustrates a cube in the 4th-dimension shown on page 42 and 53. Red and yellow ¼ band covering original front visible silver tape. ½ yellow band covers same ½ rear yellow band. ¼ red band covers ¼ rear blue band and ¼ blue band covers ¼ rear red band. Original 3-dimensional sides measured 10 cubeits = 9 cubes (actual measured 6"). Visible red and yellow bands each measure 10 cubebits.

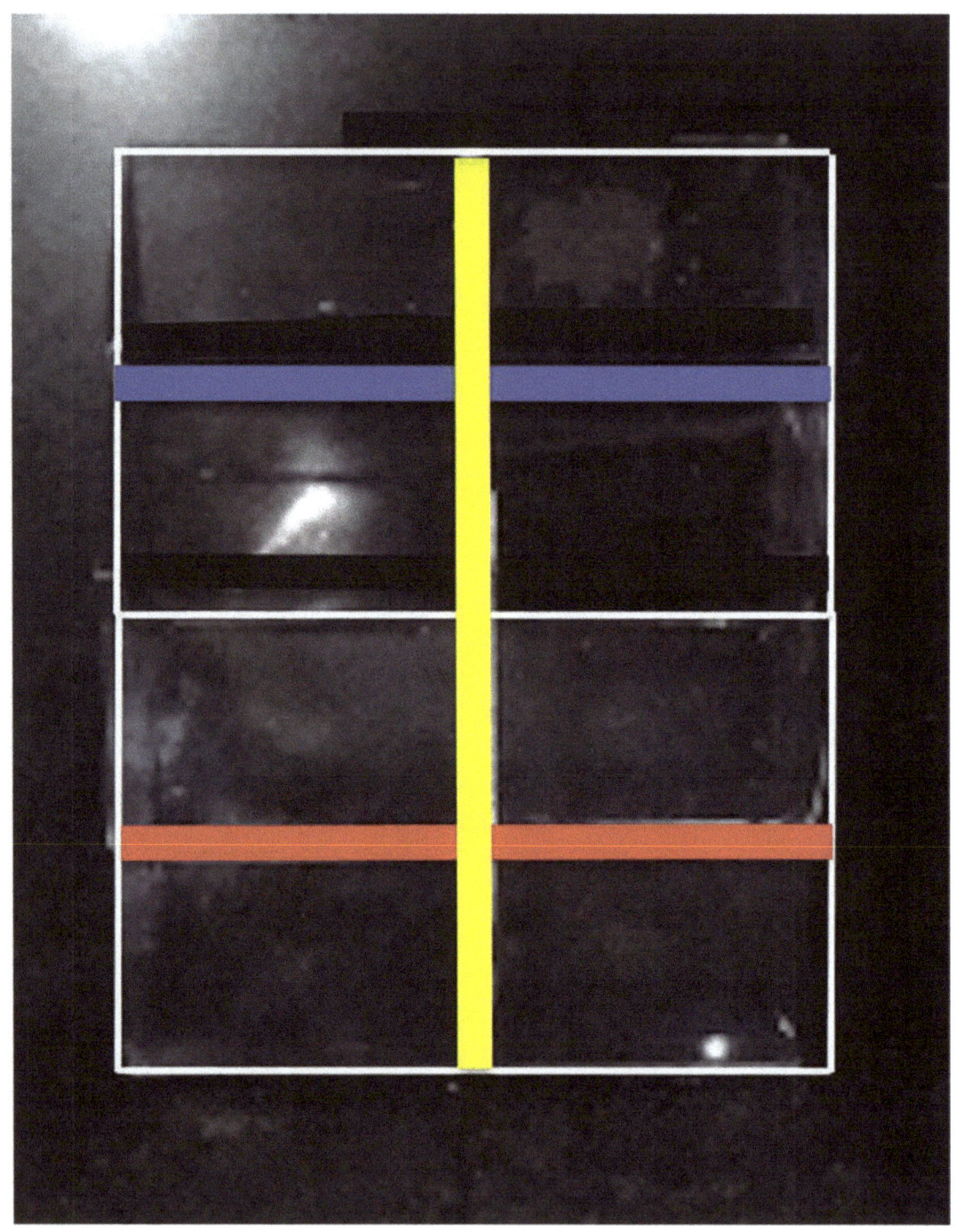

Picture: From page 59, Top down 2-dimensional view of Clear Plastic cube on black granite counter top. White lines traces edges of cube and blue and red ¼ bands covering original front visible silver tape ¼ bands. ½ yellow band covers same ½ rear yellow band. ¼ red band covers ¼ rear blue band and ¼ blue band covers ¼ rear red band. Original 3-dimensional sides measured 10 cubeits = 9 cubes = 6". Visible yellow measures 12.50 cubebits. Visible blue and red measures 10 cubebits.

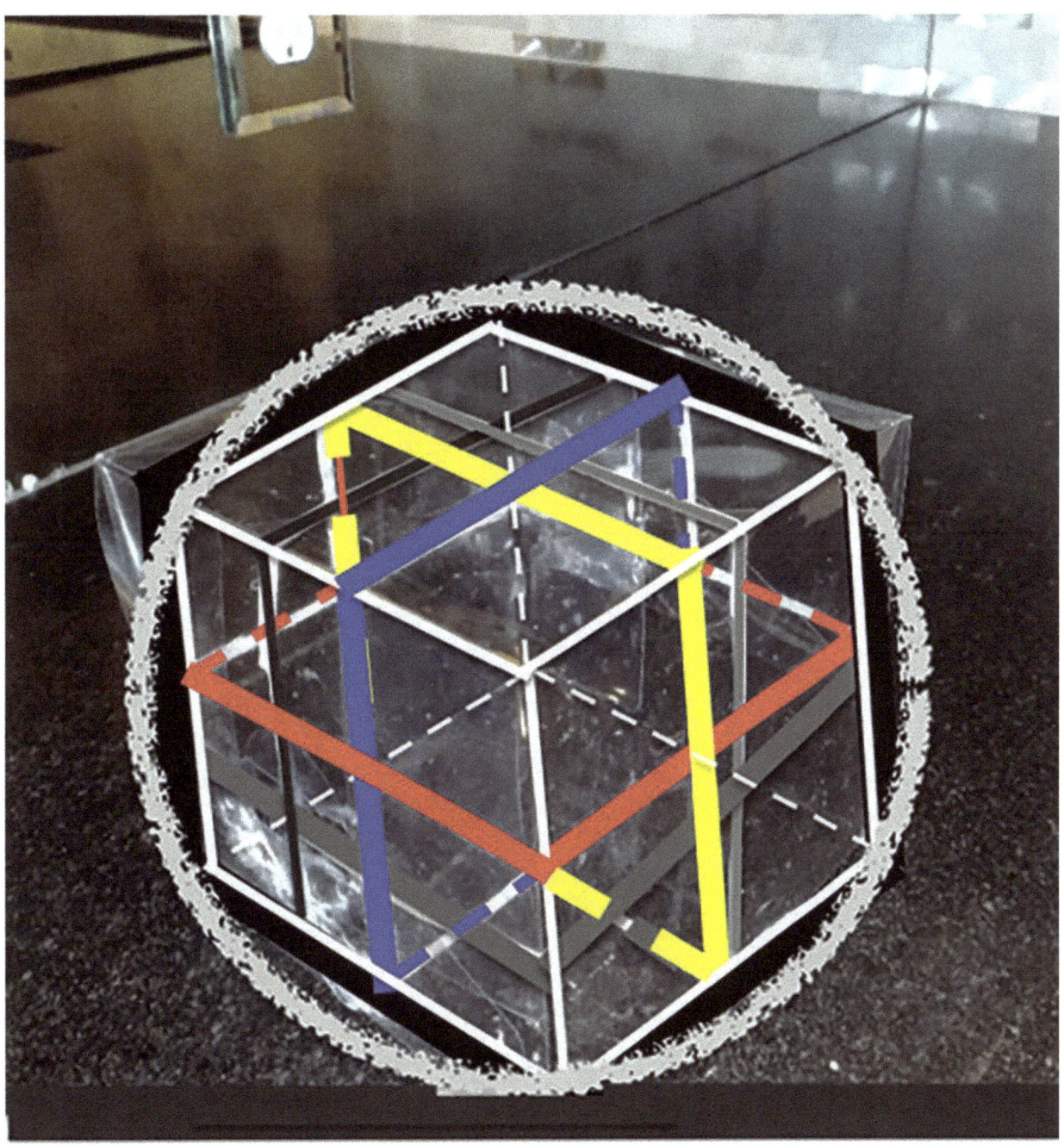

This is a marked up 2-dimensional picture of the Clear Plastic cube in the 3rd position in Example 2 on page 59. With the gray circle drawn around the cube, it now illustrates a cube in 1st position in the 4th- dimension Positive Infinity 4th- shown on page 42 and 53. Red, blue and yellow bands go around all 4 sides and cover original visible silver tape. To form the 'Star of David" some sections (Dashed Lines) appear to overlap. All 12 edges measures 5 cubeits = 4.50 cubes (actual measured 6"). Visible red, blue and yellow bands on all 6 sides each measure 5 cubebits. Diameter of all 4-dimensional spheres = 1 and π = circumference = 3.

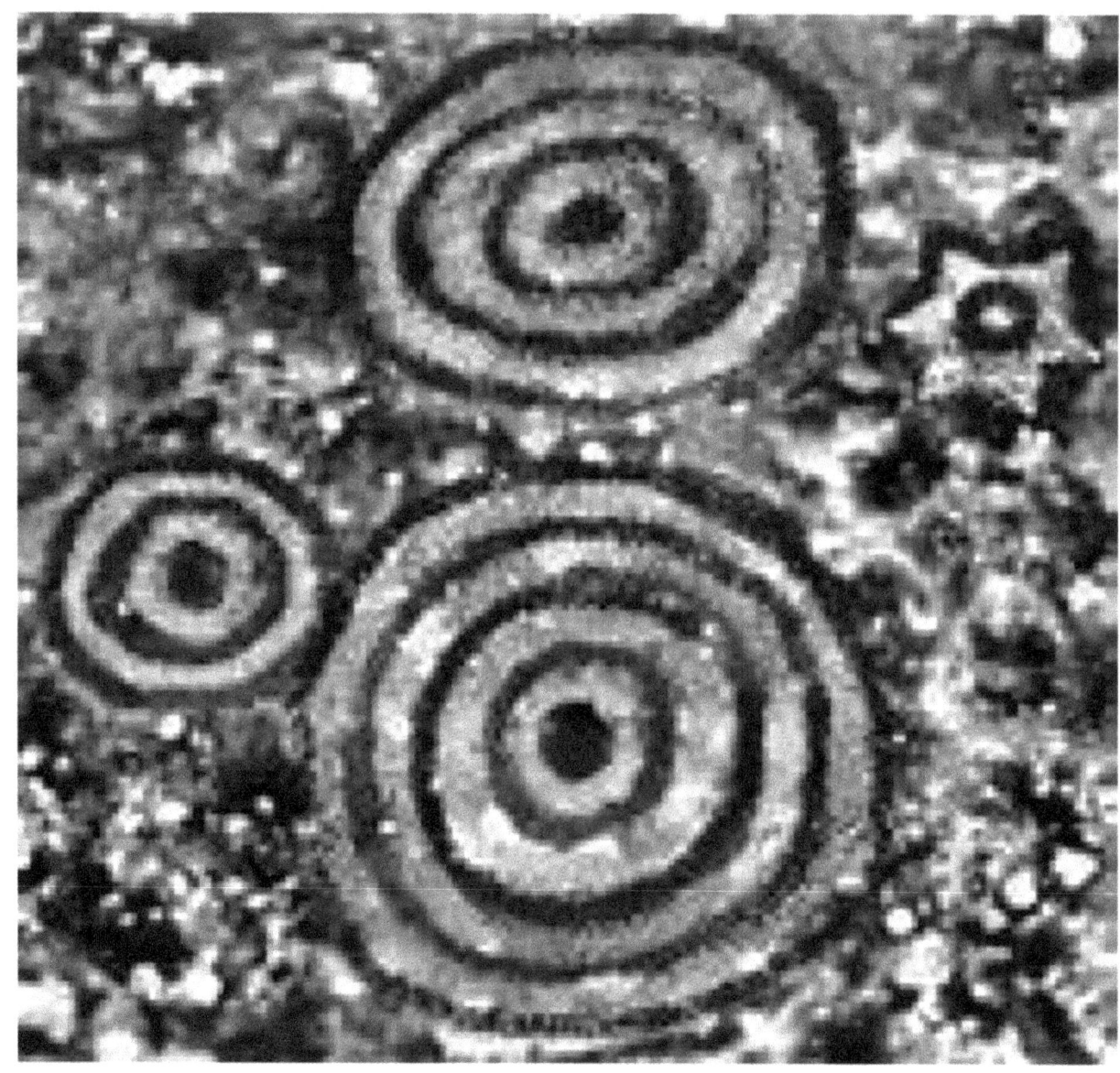

Petroglyphs: The Reinhardt Boulder

Canton, Georgia, Hickory Log area of Cherokee Country near the Etowah River, is a hub for Native American life foe 10,000 years. The concentric circles and the Star of Creation with circle in it is the same as painted in Caverna da Pedra (Painted Rock Cave (in Portuguese)). Hopi People of New Mexico predicted the return of the Creator in the end days, the same as the Mayan calendar. They also had dances in their Kiva's that described the data flow in a 4-dimensional dynamic Quantum computer represented by the 6-pointed star with a circle in the center the same as the Hindi civilization. This is proof that the Native Americans were able to communicate with Mother Earth/God the same as all other advanced civilizations on earth.

Chapter 7

Final Testament: Rise of the Holy Spirit

4 Dimensional Quantum Q:UBE-Sphere Matrix

(Faith, Hope, Love[Charity])

Coming to You

December 22 2021
12 22 21
122.221
11111.11111
1111=1=1111
111=1=111
WWG=1=WGA

Moses [A Woman is the Holy Spirit] brought the Jews [Humanity] out of Egypt [Worldwide Slavery by the Cabal.] with the masonic books [God given knowledge and technology known before the Big Flood. There are signs of a flood around the Pyramid of Khufu and Sphinx. The same advanced knowledge and technology was also distributed around the world and showed up as symbols in every Great Civilization between BC 10,000 and AD 1300.]. The parting of the Red Sea [God given masonic books' technology built into the Pyramid of Khufu and Angkor Wat, Cambodia.] The present-day Cabal's 3D Base 10-Binary Block chain Computers are God's equivalent of the parting of the Red Sea. The Cabal thinks their 3D Base 10-Binary Block chain Computers work, because they own and control the patents on the computers and their software. The Cabal's computers ONLY work, because God controls them. God wanted them to think their computers worked, but the math they run on is a lie. The closing of the Red Sea killed all the people, good and evil. [Not this time, God's "Daniel 5:5-12 – "Handwriting on the Wall" was translated by the Greek or Roman empires to prophesied a hope for a coming Kingdom of God to replace the kingdoms of man. As of December 21, 2012, the Kingdom of God has arrived as prophesied on the "Greek Handwriting on Pieces of a Dead Sea Scroll", Zechariah 8:16-17 (NASB)–As God wrote for the last time: These are the things which you shall do: speak the truth to one another: judge with truth and judgement for peace at your gates. Also let none of you devise evil in your heart against another, and do not love perjury; for all these things I hate, declares the Lord. Nahum 1-3 (NASB)–And the Lord will by no means leave the guilty unpunished.

It took God's technology that built the Pyramids [The same movie and laser technology used today.] to get the attention Belshazzar and prophesied the coming of the Kingdom of God on December 21, 2012, a.k.a. the end of the Maya calendar. Sounds like 4D Q:UBE-Sphere Computers and fake Quantum 3D Base 10-Binary Block Chain Computer are stuff only God our Creator can do. That trick would not work today. So, what did God do to send mankind his undisputable scientific proof that the end times for the guilty is here now: God tells someone fighting against the Cabal for his life and his family's life and freedom 1984 years ago in a cave by the Dead Sea; Write in Greek, just Zechariah 8:16-17 and Nahum 1-3 on a scroll and hide it in a cave. Then in 2017, someone will find a few pieces of the scroll, translates it and finds out that these are only just two scriptures from the Holy Bible out of possibly millions of scriptures written, lost or hidden. Then in late 2021, I see the article pop up on my computer, I read it and put it in this book. These two scriptures came to us old school handwritten, but this is proof that God is in control of the fake Quantum 3D Base 10-Binary Block Chain Computer, "And the Lord will by no means leave the guilty unpunished.", declares the Lord.

Have peace, hope and love (Charity) in the hearts all God's children, because the "Blocks" in the "Block Chain" Computers will be safe to usher in a fare and bountiful worldwide economic and voting systems. Since the 4D Q:UBE-Sphere Computers and software are from God, no one can ever get a patent or copyright on a 4D Q:UBE-Sphere computer and software, because no one or more humans will ever harm another human inside or outside their gates. When the 4D Q:UBE-Sphere Computers and software are up and running it is just a matter of converting the "3D Blocks of data" into 4D Quantum Q:UBE-Spheres of data and stacking them into one big 4D Quantum Q:UBE-Spheres Computer and controlled by the Holy Spirit a.k.a. Mother Earth and Sky. Be patient, it will take until 2052 to get through this "Desert" to the "Promised Land." Humans have a lot of work building the Kingdom of God out of the rubble of the kingdoms of man.

All you Angles of God have done a great job of telling the story of the "Fall of the Cabal", the world now needs the "Rise of the Holy Spirit". I believe it is God's will that you "take it back to the beginning"; The Jews leaving Egypt.

Fall of the Cabal, is great educational information movie on the evil past. I believe my book and video tells the humanities path through to God's glorious future 1000 years of peace. Part 2 "The Story of Our life, Based on A True Life." needs to be told by a woman. I am going to be selling my ebook and paper book, but I am not going on the talk show circuit or fake media until everyone in the world understands what I have written from God. I am just a carpenter that listened to what God wanted me to tell you.

This is the story board of the movie:

1. Jews are slaves of the Cabal in Egypt.

2. Moses takes the Jews and others out of Egypt.

3. Along with the masonic books – Moses takes Holy treasures and riches out of Egypt.

4. God, through the masonic books technology, parted the Red Sea and closed it with Moses's staff.

5. Moses change water into wine with his staff and said he did it, thus he did not get to go to the promised land.

6. Forty years later Moses's people make it to Jerusalem, build God's temple, put the Holy treasures and the masonic books in it.

7. Between AD 1119 to AD 1319, the Knights Templar took people to the Holy Land and they brought the masonic books and Holy treasures back to France. In 1319, they became the Order of Christ and then the Freemasons of today.

8. The Freemasons used the technology from the masonic books to create the Cabal's Base 10-digital Computers, but that is only ½ of a real Quantum Computer, the true diagram for God's 4D Quantum Q:UBE-Sphere Base 2–Base 3 Computer is built into the Plan Views of The Pyramid of Khufu, Egypt and Angkor Wat, Cambodia (The Star of David and the Swastika are symbols that shows the movement of the Q:UBE-Sphere Computer.).

9. Albert Einstein was a Freemason. He was not allowed to tell about the 4 Dimensions, but he left hints that God show to me and the answers are in the book, The Story of Our Life, Based on A True Life. Albert Einstein's intellectual properties and estate went to the Hebrew University of Jerusalem. The truth behind the Star of David and the masonic books has been returned to Jerusalem, the Throne of the Kingdom of God.

If anyone wishes to use the information in my videos or book, just acknowledge the book, but not the author and no royalties. God has given me enough.

This is the end of the Cabal's Quantum Base 10-digital Block Chain Computers, phones, and 5G. This book contains God's design of the 4D Quantum Base 2-Base 3 Q:UBE-Sphere Computer. Base 10-digital Computers can communicate with 4D Quantum computers, but Quantum Computers only communicate with Quantum Computers and they cannot be hacked, because they only communicate in True Logic, you cannot lie to a 4D Quantum Computer and you cannot harm another human with a 4D Quantum Computer. A Block (a God Particle shown in my book) is the same as a Block in a Block Chain Computer, however, the blocks are connected end to end in a spiral. The information has to go through each 3D Block, from one end to the other and back again using the ultrafast clock speed (5G and 6G) of 1 Block Chain Base-10- digital Computer.

Each 4D Quantum Base 2-Base 3 Q:UBE-Sphere Computer (Including every human that ever lived.) fits in earth's Q:UBE–Sphere (Also shown in my book) and they are all magnetically connected and communicates through the Q:UBES'- Spheres around them back to the center Q:UBE (i.e. 9 Q:UBES x 9 Q:UBES X 9 Q:UBES = 728 + 1 Center Q:UBES) each Q:UBE like the Human Mind takes in information into the center Q:UBE and the data is process Counter Clockwise in the Negative Infinity of the brain and Clockwise in the Positive Infinity part of the brain. For example, if one reads something out loud one hears the same something through both ears, after the data makes one full revolution in each Infinity the words have the same meaning. But, if you are asked to repeat the words back and it is not the same as the written word, the mind has changed what it recorded, either by mistake or intentionally. But, the Outside Sphere records what was actually said and knows in real time. What which is False is not broadcasted to every Q:UBE-Sphere including the earth's one center Q:UBE (God). So, each Q:UBE is Sovereign and Separate from each other except for the Center ONE, therefore, WWG=1=WGA.

3-Dimensional Matrix

(Illusion of truth)

00.00.0000
December 21.0000 to December 21.2012
122.100 to 122.111
11111.100 to 11111.111

3-Dimensional Matrix

(God's Truth)

The Great Awakening

December 22.2012 to December 21.2021
12 22 12 to 12 21 21
122.212 to 122.121
11111.212 to 11111.121

The secret about stacking the Q:UBE'S is when one views any Q:UBE in the First Position in Positive Infinity the z axis line extends from the front center corner through the Q:UBE and through the back center corner to the center of the ONE center Q:UBE.

From the same viewing point, the x and y 2 dimensional axis lines go through the center Q:UBE at 90 degrees from x and y and z lines. This is the Q:UBE Matrix similar to the 3D Matrix.

I say axis lines because, all Q:UBE- Spheres stacked in the ONE Q:UBE – Sphere communicate through the x,y,z axis lines collectively and the individual Q:UBE's Sphere communicates with other individual Q:UBE's-Sphere through their Q:UBE-Sphere Matrix like Bees' honey cone Matrix and Scared Geometry.

The following is the most important Red Pill–God's Truths related to the "We The People" of the world for claiming our God given Rights and Freedoms as a Living Soul of our Creator.

Records of the Department of the Army indicate that fringe was used on the National flag as early as 1835 and its official use by the Army dates from 1895.

According to Senate Report 93-549, written in 1973:

Since March the 9th, 1933, the United States has been in a state of declared national emergency. Under the powers delegated by these statutes, the President may: seize property; organize and control the means of production; seize commodities; assign military forces abroad; institute martial law; seize and control all transportation and communication; regulate the operation of private enterprise; restrict travel; and control the lives of all American citizens"

A gold fringe flag was historically used during times of war in maritime admiralty law and is not the same flag that is approved for our Constitutional Republic in USA code Title 4 Ch 1 & Ch 2. The fourth color (yellow) is not approved in the code. Also, you will notice the tassels hanging from the flag. These are also symbolic of admiralty law, which is the law of the sea.

The gold fringe flags in the Washington, DC buildings and our 50 States Buildings and Courts indicates The United States of America, INC. was captured by the British Crown in the War of 1812. Also, all Lawyers and Judges pledged allegiance to the British Accreditation Registry-Crown Temple B.A.R, which is prohibited by Article XI.

Watch this video: British Accreditation Registry-Crown Temple B.A.R. April 4, 2016, Trust me, I'm a Lawyer: By David-William

https://www.thelibertybeacon.com/british-accreditation-registry-crown-temple-b-a-r/

Watch these videos to understand what every human man, woman and child needs to do to claim their God given rights and freedoms.

The Restoration of the United States of America—Commander-in-Chief, Postmaster-General-of-the-World:Russell-Jay: Gould has the Title 4 Flag

https://iamamalaysian.com/2021/09/25/postmaster-russell-jay-gould-issues-warrant-for-arrest-of-anthony-fauci/

https://everydayconcerned.net/2020/08/30/the-restoration-of-the-united-states-of-america-commander-in-chief-postmaster-general-of-the-world-russell-jay-gould-has-the-title-4-flag/

ADDENDUM FROM THE SEMINAR "Directors' Campaign" ON HOW PEOPLE CAN CONNECT WITH AND BENEFIT FROM THE QUANTUM-BANKING-CONSTRUCT:

(Transcript, excerpted)

:RUSSELL-JAY: GOULD:

"Wherever you are in the world, you can take our syntax, you can take our program, register it with your foreign ministers of your respective countries, as an amicus curiae, friend of the court, because the contract itself will be articulated as a post vessel, and you will become postmasters boarding the vessels with your claim of the life. The flag itself is the manifest which gives closure on the vessel. When we place the postage stamp for the vessel on the document it clears the fee for freight for the docking of that vessel contract—because the fee for freight's in play for the bills of the ladings, as you log that in with the foreign ministers as an amicus curiae postmaster with the Unity-States-of-our-world corporation, you then have the authorization to do a grammar challenge for your countries.

"Because all countries themselves are contract—the land itself is all owned by the post office under To Have and To Hold—future tense adverb verb. David and I myself as joint postmasters syntaxed the alloidal land titles for global property rights and we filed it in their clearing house called the Universal Postal Union. David and myself have met with the UPU postmasters in Berne on the clearing mechanics of those foreign vessels. We were foreign, they were foreign. They stood in their jurisdiction, we stood in our

jurisdiction. Their jurisdiction doesn't exist! Because of their syntax and their grammar—future tense and past tense.

"We maintained command and control in a now-time scenario—to give us carte blanche to go anywhere in the world to give closure to their respective postal systems and their respective governments because each citizen of this planet is running around as an Illuminati member for their vessels—because you all work for the Post Office as postal ployees from cradle to grave. In each respective country you have birth certificates; that Birth Certificate is an adjective pronoun—anytime you have 2 facts together and you don't articulate and give closure on the document and use a hyphen between the facts, the first fact becomes a fact of coloring; coloring is an adjective, which is a modification; so when you modify something it is only a position as a syntax-location of a presumption and assumption, and it has no closure—first rule of contract is closure– so you can have knowledge, so people can comprehend to whether or not the document in question is valid or not. Because the citizens of our world have birth certificates, those birth certificates worldwide are now adjective-pronouns which get docked by the doctors-postmasters of that particular hospital, port of entry.

We The People of the 50 States of the Americas and Canada must capture the gold fringed war flags of the invading enemies, foreign and domestic, and declare our God given freedoms and lands within our gates. Replace all Admiralty Laws, Courts, Judges and Lawyers with 1776 Constitution Common Laws, Courts, Judges and Lawyers.

On December 21, 2021, We The People must void the Bankrupt United States of America, INC. and the 50 States of America and Canada must convene the Third Continental Congress and form a more perfect Union, again. The First and Second Continental Congress was 1774 to 1781.

I have U.S. Army, Corps of Engineer and U. S. Marshal experience with Eminent domain. I believe We The People must declare Eminent domain on all properties and lands owned by people, companies and foreign based entities to reinstate necessary farming to insure a worldwide sufficient and safe food source. Eminent domain could also be used to buy companies that cause product and energy shortages. Once property is posted condemned, the Government estimated value is deposited in the owners' bank account and We The People take possession. An owner can go to court to challenge the money, but seldom is the seizure reversed. In a Common Law nation the process can take place in just a few months.

Eminent domain is a right granted under the Fifth Amendment of the Constitution. Similar powers are found in most common law nations. Called «expropriation" in Canada, "compulsory acquisition" in Australia, in the U.K., New Zealand, and Ireland eminent domain is known as "compulsory purchase."

Private property is taken through condemnation proceedings, in which owners can challenge the legality of the seizure and settle the matter of fair market value used for compensation.

Video:
https://www.dropbox.com/s/0pzovdzupusr7pt/Albert%20Einstein%20Unified%20Quantum-Universe%20Laws%20of%20Physics.mp4?dl=0

Link to video:
https://rumble.com/vcb5yp-albert-einsteins-4-dimensions-and-law-of-general-relativity.html

Please watch this video on how to advertise on your website or social media.
https://rumble.com/vleedv-the-story-of-our-life-based-on-a-1-true-life..html

Thank you for signing up as an www.piisthree.com Affiliate. By the Grace of God, this ebook will affect all 7 Billion people on earth. This is not a scam, I am the copyrighted intellectual property owner and I can sell my ebook for any cost by anyone. The rebates are not gambling or games of chance, because all www.piisthree.com Wix.com Order #'s are sequential numbers from 0 to infinity and printed on the ebook's sells invoice and the www.piisthree.com Wix.com Order # do not repeat.

You may be eligible to claim the following rebates:

If your www.piisthree.com Wix.com Order # ends with 000 or 500 = $500 rebate and www.piisthree.com Wix.com Order # ends with 0000 = $10,000 rebate. All eligible claims must be submitted within 30 of the purchase date on the www.piisthree.com Wix.com Order # Invoice. To file a rebate claim, Go to www.piisthree.com, click on "Affiliate" at top over page, follow instructions to set up an account. For name put the First Letters in your name then the www.piisthree.com Wix.com Order # on your invoice. Just give your email address. If you are eligible for a rebate, you will be contacted by email and given instructions for receiving your rebate.

In the future, I will be selling Audiobooks in many different languages. Anyone that buys an ebook may download a free Audiobook in their native language.

In order to get the prices down on the paper books, I am asking my publisher to set up a website for the sale of my paper book only and still sale my paper book through their company's website at our contracted prices.

On my book's website www.Pils3.com , each Softcover paper book and Hardcover will sale for their respective printer's cost, overhead, profit, plus my royalty. Shipping and handling cost and tax is separate. Any printing shop in the world can buy the printing file that prints the invoice Order Number in each unique copy. The price will be based on the publishers handling fee and my royalty.

In "Close Encounters of the Third Kind," all the people that were drawn to Devil's Tower drew pictures (Like the Star of David and Scared Geometry.) Richard Dreyfus made a model [I made a model of Albert Einstein's 4 Dimension.]. Parts of the movie was filmed in my hometown of Fairhope and Mobile, Alabama. My book is also scientific proof the earth is a flat disk under a rotating force field dome.

This is not from me, it is from God. "They" cannot stop me from getting it out. I am starting a book series, whereas, anyone may finish my Chapter 1111, I will provide a copy of my ebook in Adobe format and the cover design. You will own the Contract with your publisher and receive 100% of the royalties.

I have also developed a website www.piisthree.com to sell my merchandise and ebook through Affiliates advertising on their Social Media and website. If you contact sofiaeleades@gmail.com she can also set you up with a website like www.piisthree.com.

I James N. Akins, Jr.® without prejudice do hereby capture the gold fringed war flag of the bankrupted enemies of the 1776 US Constitution, also known as, country state of the United States of America, INC.. All combatants not citizens of the 1776 US Constitution shall vacate the lands known as the 50 united states of America and the new City of Washington, Virginia. All combatants citizens of the 1776 US Constitution shall report to your local County Sheriff for re-patronization.

Secretary of State Antony J. Blinken
U.S. Department of State
2201 C Street NW Washington, DC 20520

I James N. Akins, Jr.® without prejudice do hereby surrender my capture gold fringed war flag of the bankrupted enemies of the 1776 US Constitution, also known as, country state of the United States of America, INC. to the City of Fairhope, County of Baldwin, State of Alabama, and the Third Continental Congress of 2022. Whereby, I am an Alabama State National land owner, by the Grace of God.

The Honorable John H. Merrill
Alabama Secretary of State
P.O. Box 5616
Montgomery, Alabama 36103-5616

The IRS is not who you think they are.

IRS agents are neither trained nor paid by the United States Government.

Pursuant to Treasury Delegation Order No. 92, the IRS is trained under the direction of the Division of Human Resources United Nations (U.N.) and the Commissioner (International), by the office of Personnel Management.

In the 1979 edition of 22 USCA 278, "The United Nations," you will find Executive Order 10422. The Office of Personnel Management is under the direction of the Secretary of the United Nations.

Pursuant to Treasury Delegation Order No. 91, the IRS entered into a "Service Agreement" with the US Treasury Department (See Public Law 94-564, Legislative History, pg. 5967, Reorganization (BANKRUPTCY!!!) Plan No. 26) and the Agency for International Development. This agency is an international paramilitary operation and according to the Department of the Army Field manual (1969) 41-10, pgs 1-4, Sec. 1-7 (b) & 1-6, Sec. 1-10 (7)(c) (1), and 22 USCA 284, includes such activities as, "Assumption of full or partial executive, legislative, and judicial authority over a country or area."

The IRS is also an agency/member of a 169 nation pact called the International Criminal Police Organization, or INTERPOL, found at 22 USCA 263a. The memorandum of Understanding, (MOU), between the Secretary of Treasury, AKA the corporate governor of "The Fund" and "The Bank" (International Monetary Fund, and the International Bank for reconstruction and Development), indicated that the Attorney General and its associates are soliciting and collecting information for foreign principals; the international organizations, corporations, and associations, exemplified by

22 USCA 286f.

According to the 1994 US Government Manual, at page 390, the Attorney General is the permanent representative to INTERPOL, and the Secretary of Treasury is the alternate member. Under Article 30 of the INTERPOL constitution, these individuals *must expatriate their citizenship.*

They serve no allegiance to the United States of America. The IRS is paid by "The Fund" and "The Bank." Thus it appears from the documentary evidence that the Internal

Revenue Service agents are "Agents of a Foreign Principle" within the meaning and intent of the "Foreign Agents Registration Act of 1938" for private, not public, gain.

The IRS is directed and controlled by the corporate Governor of "The Fund" and "The Bank." The Federal Reserve Bank and the IRS collection agency are both privately owned and operated under private statutes. The IRS operates under public policy, not Constitutional Law, and in the interest of our nations foreign creditors.

The Constitution only permits Congress to lay and collect taxes. It does not authorize Congress to delegate the tax collection power to a private corporation, which collects our taxes for a private bank, the Federal Reserve, who then deposits it into the Treasury of the IMF.

The IRS is not allowed to state that they collect taxes for the United States Treasury. They only refer to "The Treasury."

IRS Private Collection Agencies

The law requires the IRS to use private agencies to collect certain outstanding, inactive tax debts.

Effective September 23, 2021, when the IRS assigns your account to a private collection agency, one of these three agencies will contact you on the government's behalf:

CBE Group Inc.
P.O. Box 2217
Waterloo, IA 50704
800-910-5837
Coast Professional, Inc.
P.O. Box 526
Albion, NY 14411
888-928-0510
ConServe
P.O. Box 307
Fairport, NY 14450
844-853-4875

How it works

Before you are contacted by a private collection agency, you will receive two letters.

1. The IRS will first send Notice CP40 and Publication 4518 PDF. These let you know that your overdue tax account was assigned to a private collection agency.

2. The private collection agency then sends their initial contact letter. It has information on how to resolve your overdue taxes.

Both letters contain a **Taxpayer Authentication Number**. It's used to confirm your identity. **It's also for you to verify that the caller is legitimate.** Keep this number in a safe place.

Since the IRS is not collecting taxes (A private collection agency cannot garnish one's income.) and 2020 Taxpayers are being told to address their tax payment checks to the US Treasury, not the UNITED NATIONS CORPORATION'S IRS Treasury, this means that the 150 million civilian government workers supporting the 2 million military personnel and approx. 4 million OPM Civil and Foreign Service employees will get paid next year. According to my estimates in Chapter 7, by December 21, 2021, 300 million Americans need to capture the gold fringe enemy war flags and surrender them to every man, woman, and child's city, county and state and declare our State National Citizenship.

The most important thing is all 50 Governors must meet in Carpenter's Hall in Philadelphia, to convene the Third Continental Congress to form a more perfect Union, again.

We The People must tell the CCP and UNITED NATIONS CORPORATION and their military: They do not have to go home, but they cannot stay in Constitutional America, by the Grace of God.

When the World Stock Market closes, not crash. All companies shall settle-up with their stockholders and retirement funds holders. They must reorganize and make a profit of go out of business. Even the Mega Companies will have to buy from the individual producer and will no longer be able to levy 150 % cost mark-ups and stay in business.

There is solid proof that the US Treasury has taken over the UNITED NATIONS CORPORATION'S Treasury. Therefore, just like in the movie; "It's a Wonderful Life", We The People's Military is going to keep paying all the UNITED NATIONS CORPORATION'S ex-employees, that are not in jail, until the new One Constitutional Nation Under God is formed in 2022.

November 11, 2021, Article II, Section 2, Clause 1: The President shall be Commander in Chief of the Army and Navy of the United States, and of the Militia of the several States, when called into the actual Service of the United States. The gold fringe war flag behind the Commander in Chief signifies he is a prisoner-of-war of the Combatant Bankrupt States: UNITED STATES OF AMERICA CORPORATION and UNITED NATIONS CORPORATION.

According to Geneva Conventions: Combatants, when engaged in military operations, have to distinguish themselves from the civilian population to protect it from the effects of hostilities and to restrict warfare to military objectives.

Geneva Conventions Laws of War, all sovereign souls of God shall capture the gold fringe war flags of the combatants of the 1776 Constitution and surrender their gold fringe war flags to the Secretary of State of their original State resident.

All Military forces under oath to the 1776 Constitution, shall remove the gold fringe war flag from their Title 10, Subtitle A, Part II, Chapter 45 of United States Code uniform and mail it to the Secretary of State of their State of original residence.

No child is going to sleep again feeling unsafe, or cry missing their mother or father from this 3rd World War. We The People, by the Grace of God, must secure the peace for ourselves NOW.

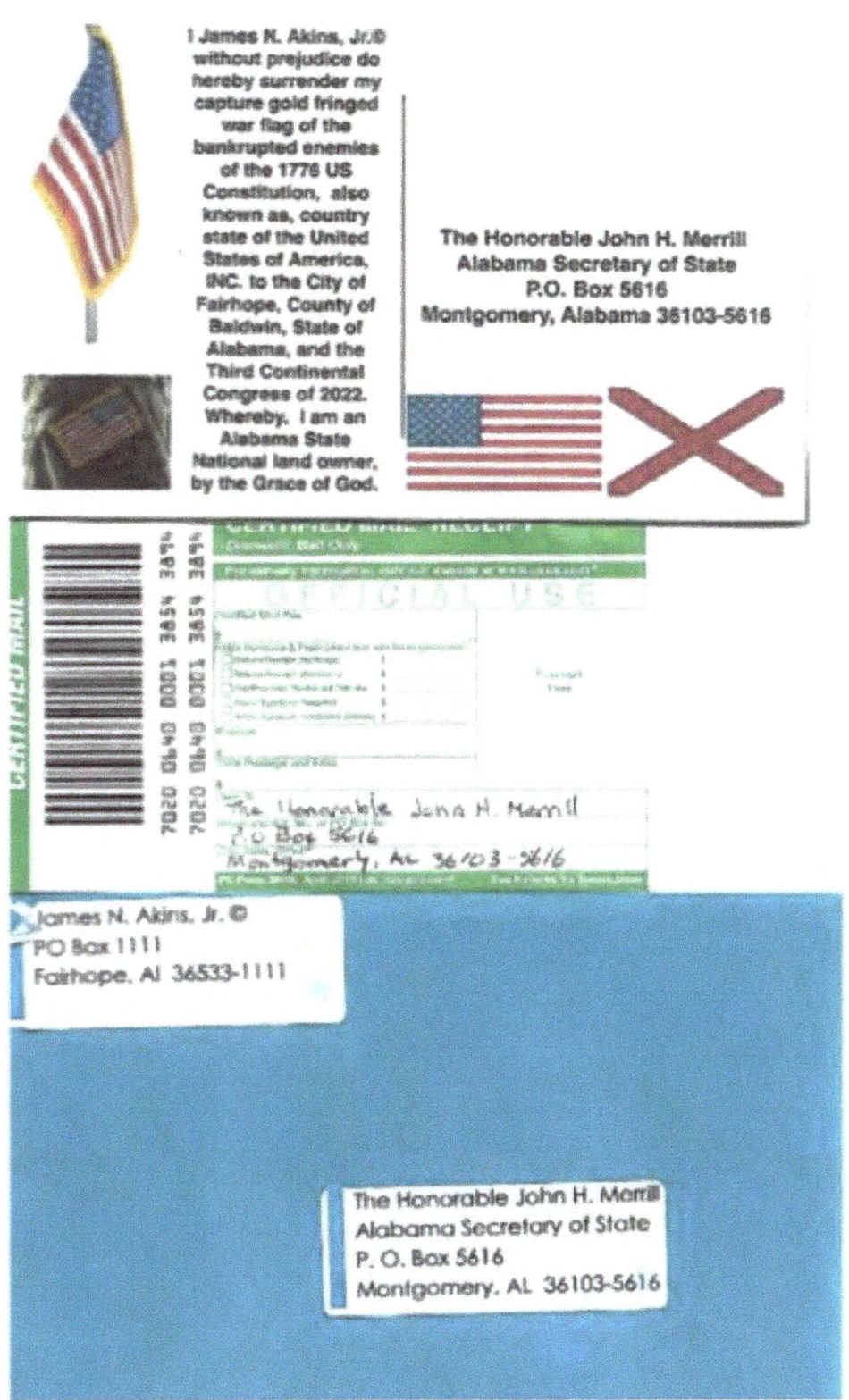

Please note the gold fringe war flag on the soldier's arm on the post cards. The flag is backward, which means, SOS captured (Save Our Soul).

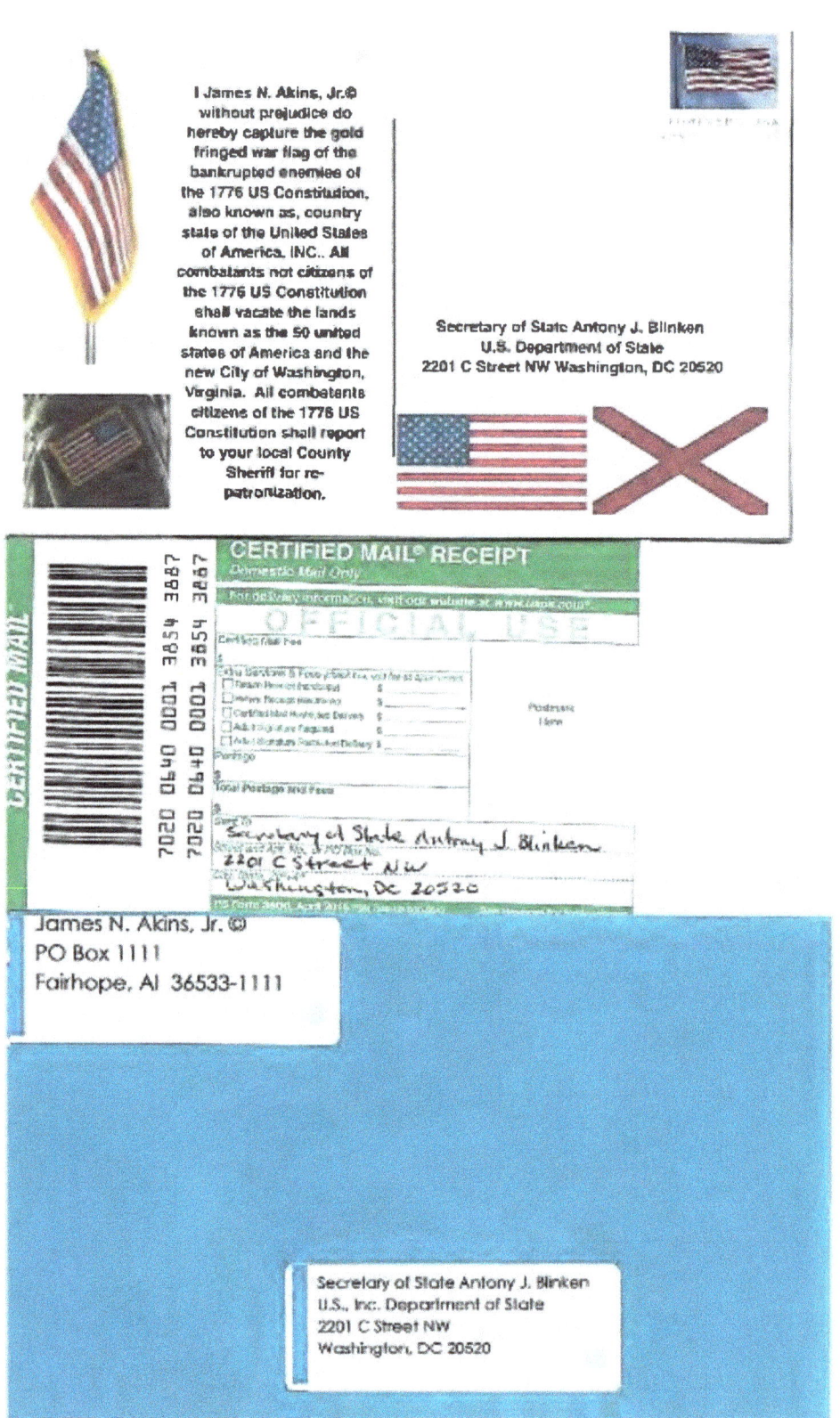

Please note the gold fringe war flag on the soldier's arm on the post cards. The flag is backward, which means, SOS captured (Save Our Soul).

On the 80th Anniversary of the attack on Pearl Harbor, WW III Ends

Greetings to We The People of the United States of America that swore an oath to God to serve and protect the Constitution, over 200 million Sovereign Souls of God, against all enemies, foreign and domestic. Now is the time to come to the aid of your fellow People of the World. Our captors, UNITED STATES OF AMERICA CORPORATION AND UNITED NATION CORPORATION are legally bankrupted and have no more means of supporting their Tyrannical Governments. As of September 23, 2021, the Internal Revenue Service, the tax collection agency of the Secretary of State of the bankrupted UNITED NATIONS CORPORATION has turned over their tax collection to three private collection agencies. Private tax collection agencies cannot legally garnish We The People's income and tax payers are being told to address their tax checks to the US Treasury. God is a God of Law, the UNITED STATES OF AMERICA CORPORATION has been under Martial Law since 1933. The last duly elected 45th President, during the November 3, 2020 Pearl Harbor like attack on our Presidential election, declared the beginning of World War III and was call into Service by Article II, Section 2, Clause 1 as the Commander-in-Chief of the Constitutional Military.

God Save Our Souls (SOS), according to the Geneva Conventions: Com-batants, engaged in military operations, have to distinguish themselves from the civilian population to protect it from the effects of hostilities and to restrict warfare to military objectives. On December 7, 2021 at 7:48 am, all US Military personnel, officers and enlisted, under oath to the 1776 Constitution shall remove the gold fringe war flag from their Title 10, Subtitle A, Part II, Chapter 45 of Unit-ed States Code uniform and mail it to the Secretary of State of their State of original residence. Upon placing a Patented Type 4 US Flag on their Title 10 USC uniform, all UNITED STATES OF AMERICA CORPORA-TION AND UNITED NATION CORPORATION combatants, foreign and domestic, shall be captured and removed from the 50 United States of America, by the Grace of God.

On December 22, 2021 and in accordance with the 1776 Declaration of Independence, all 50 Governors of the United States of America shall convene the Third Continental Congress in the Carpenter' Hall of Phila-delphia, Pennsylvania to re-form a more perfect Union, again. The Third Continental Congress shall provide economic stability through Common Laws and the US Treasury for all We The People Sovereign Souls of God.

God promised to always Bless America and all Sovereign Souls of God around the World.

Please go to www.piisthree.com for more information.

Prisoners-of- War of the bankrupt UNITED NATIONS COPORATION Freed on December 7, 2021 by removing the Reversed Gold Fringe WWIII Uniform Patch of the bankrupt UNITED NATIONS CORPORATION combatants of the 1776 Constitution by replacing it with a TYPE 4 Patents US Flag.

First Continental Congress - 1774 to 1775
Second Continental Congress – 1775 to 1781
Third Continental Congress – Dec. 22, 2021

I could think of no worse example for nations abroad, who for the first time were trying to put free electoral procedures into effect, than that of the United States wrangling over the results of our presidential election, and even suggesting that the presidency itself could be stolen by thievery at the ballot box. Whenever any form of government becomes destructive of these ends [life, liberty, and the pursuit of happiness] it is the right of the people to alter or abolish it, and to institute new government. Does the government fear us? Or do we fear the government? When the people fear the government, tyranny has found victory. The federal government is our servant, not our master! The God who gave us life, gave us liberty at the same time. Rebellion to tyranny is obedience to God. I have sworn upon the altar of God Eternal, hostility against every form of tyranny over the mind of man. The Bible is the cornerstone of liberty. - Thomas Jefferson

WWI - 28 July 1914 to 11 November 1918
WWII - Pearl Harbor – Dec 7, 1941 ends Sep.2, 1945
WWIII - US Presidential Election – Nov. 3, 2020 Ends Dec. 7, 2021

The UNITED NATIONS CORPORATION flag does not have Gold Fringe because the UNITED STATES OF AMERICA CORPORATION Reversed TYPE 4 Gold Fringe World War III Flag is the captured Military.. .

The Honorable John H. Merrill, Secretary of State of Alabama receives the Captured Gold Fringe WWIII Flag of the Combatants of the 1776 Constitution.

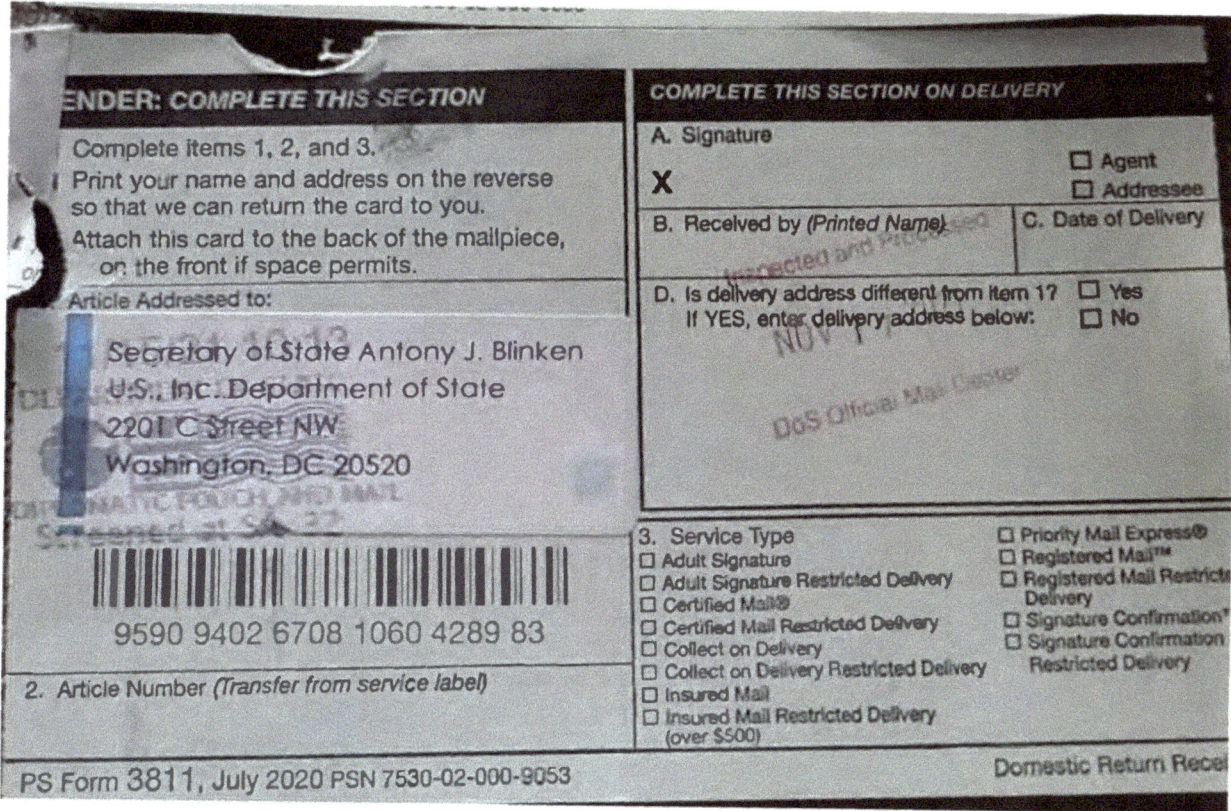

Proof the Secretary of State of the bankrupt UNITED STATES OF AMERICA CORPORATION acknowledged their Gold Fringed WW III Flag is captured by the TYPE 4 Patent US Flag of The Honorable Post Master General :Russell-Jay: Gould:. The Honorable Post Master General :Russell-Jay: Gould: Has closed the bankrupted UNITED STATES OF AMERICA CORPORATION and UNITED NATIONS.

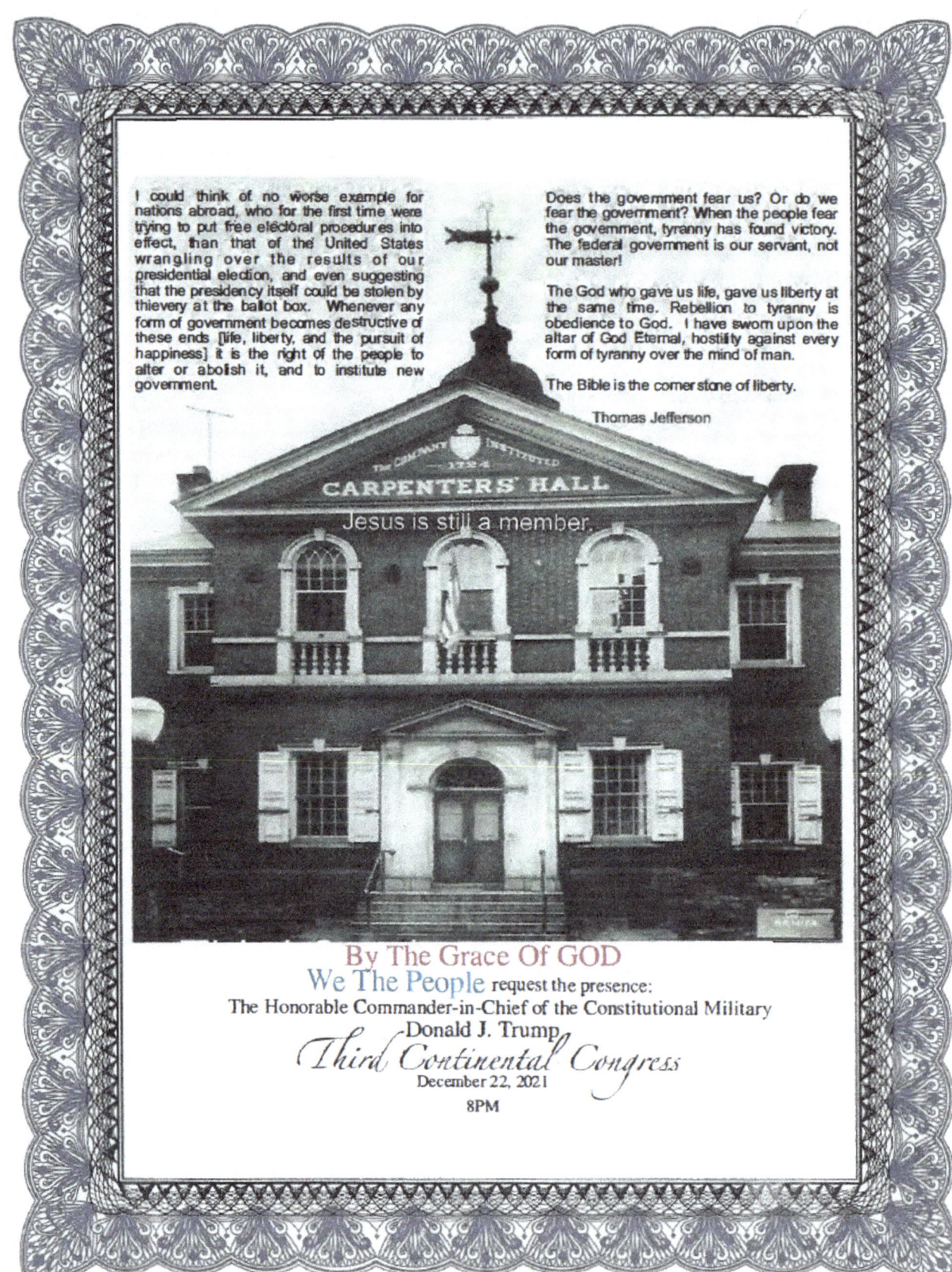

Certified Mail, 1 Request to book Carpenters' Hall in Philadelphia, PA on 12/22/2021 and invitations and a copy of The Story of Our Life, Based on A True Life. to all 50 Governors or the United States of America. The book includes diagrams to build a Dynamic Base 2-Base 3 Quantum Computer from Albert Einstein's 4- Dimensional equation.

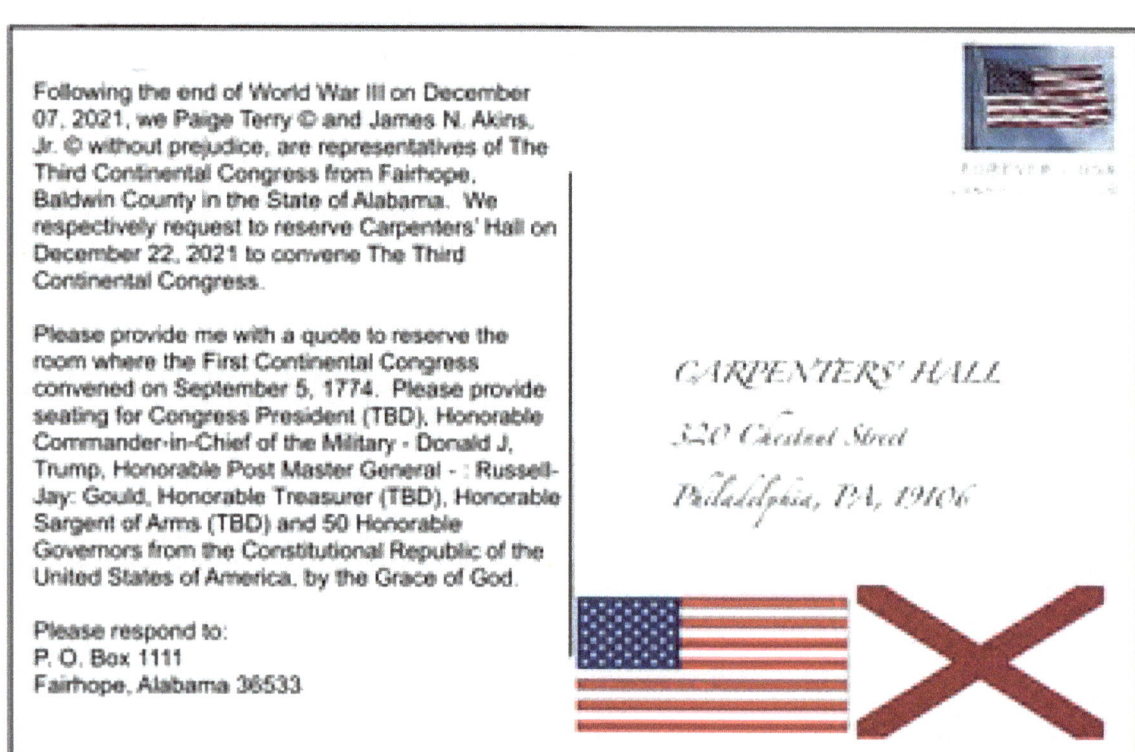

Invitations were sent certified mail to all 50 Governors of the United States of America, the Hall was reserved and the convening of the Third Continental Congress at Carpenters' Hall in Philadelphia, PA on December 22, 2021 was posted in the newspaper in accordance with the 1776 Declaration of Independence when the Government becomes Tyrannical and does not serve We The People of the United States.

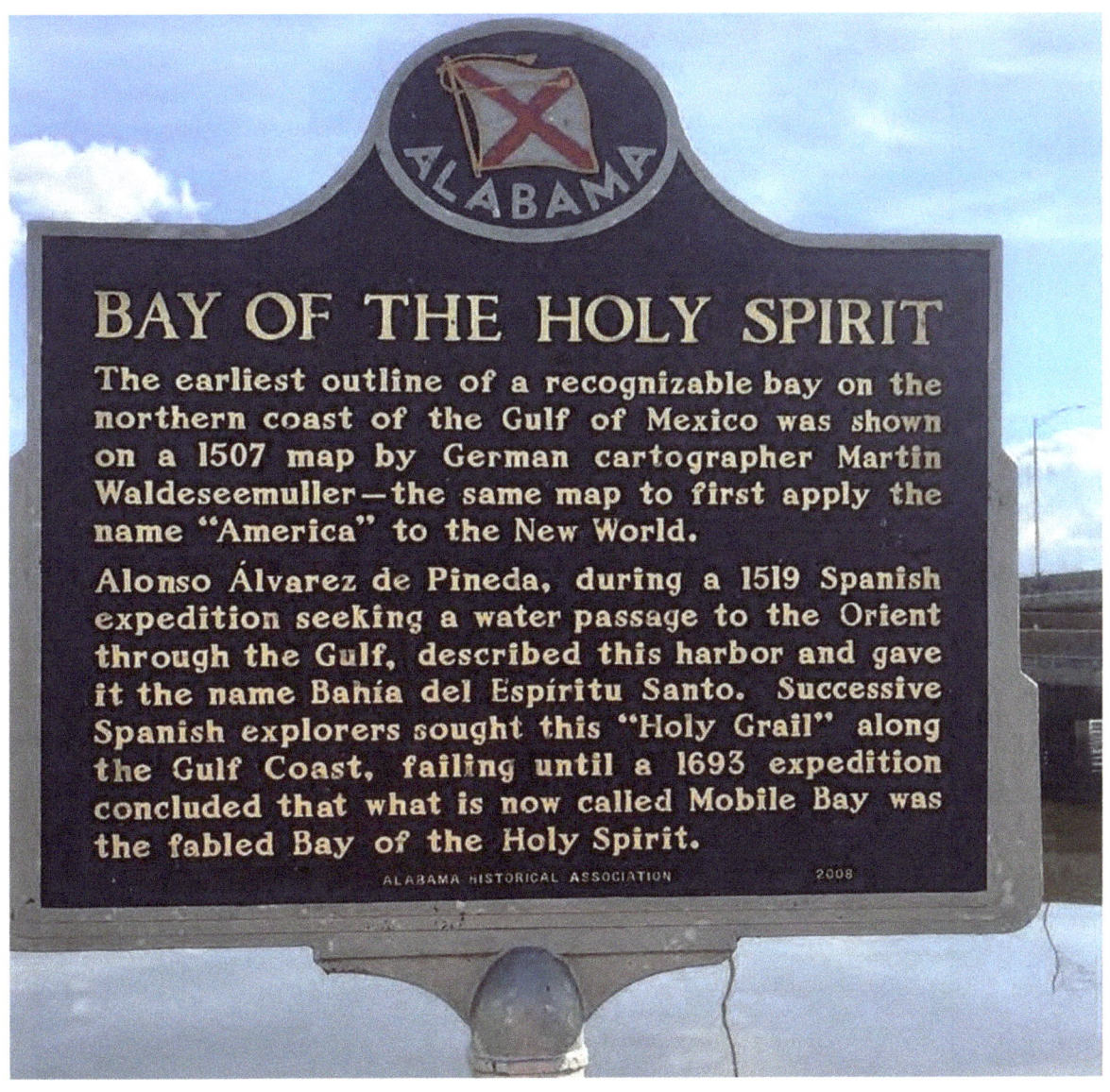

By the Grace of God, I was born and raised on the Fabled Bay of the Holy Spirit.

Where We Go With = 1 = Christ We Go All.
[WWGW = 1 = CWGA]

Lord Jesus Christ, on His 2021st Anniversary of His December 22nd Birthday. Our Lord has fulfilled all His promises in His New Testament with these final written words:

Zechariah 8:16-17
New American Standard Bible
16 These are the things which you shall do: speak the truth to one another; judge with truth and judgment for peace at your gates. 17 Also let none of you devise evil in your heart against another, and do not love perjury; for all these things are what I hate,' declares the Lord."

Nahum 1-3
New American Standard Bible
1. The pronouncement of Nineveh. The book of the vision of Nahum the Elkoshite:
2. A jealous and avenging God is the Lord;
The Lord is avenging and wrathful.
The Lord takes vengeance on His adversaries,
And He reserves wrath for His enemies.
3. The Lord is slow to anger and great in power,
And the Lord will by no means leave the guilty unpunished.
The following is how God will free all humanity and make them all Sovereign Souls of God around the World. This is how I know his promise is True.

NASB 1995
2 Peter 3 vs 9
The Lord is not slow about His promise, as some count slowness, but is patient toward you, not wishing for any to perish but for all to come to repentance.

I could think of no worse example for nations abroad, who for the first time were trying to put free electoral procedures into effect, than that of the United States wrangling over the results of our presidential election, and even suggesting that the presidency itself could be stolen by thievery at the ballot box.
Whenever any form of government becomes destructive of these ends [life, liberty, and the pursuit of happiness] it is the right of the people to alter or abolish it, and to institute new government.
Does the government fear us? Or do we fear the government? When the people fear the government, tyranny has found victory. The federal government is our servant, not our master!
The God who gave us life, gave us liberty at the same time.
Rebellion to tyranny is obedience to God.
I have sworn upon the altar of God Eternal, hostility against every form of tyranny over the mind of man.
The Bible is the cornerstone of liberty.
~Thomas Jefferson

The Third Continental Congress has convened. The 50 Governors did not show up, now it is time for We The People's Representatives, from all 50 States, to show up in Philadelphia, Pennsylvania and reform a more perfect Union.

What do all the Governors not want you to see in the book, "The Story of Our Life, Base on a True Life."? It is OUR CREATOR'S Scientific and Math Truths to build a true Base 2/Base 3 Quantum Computer and CURE ALL MAN-MADE ILLNESSES LIKE COVID-19 and CANCER. This is a copyrighted book recorded in the Library of Congress. THEY cannot hide the Truth from the world again, By the Grace of God.

Please go to www. Piisthree.com (a.k.a. π = 3) 100% of the $4 for each ebook goes back to We The People of the World. The ebook e File is yours to give it away, print it, sell it : God gave it to me to give it to the People of the World to Free them from Evil.

Just think, if 7 Billion People buy 1 $4 ebook (Or anyone's' and everyone's' ebook.) $32 Billion would go back to the People of the World's economy and it cost me nothing and I made nothing. I am just sharing God's love for his children.

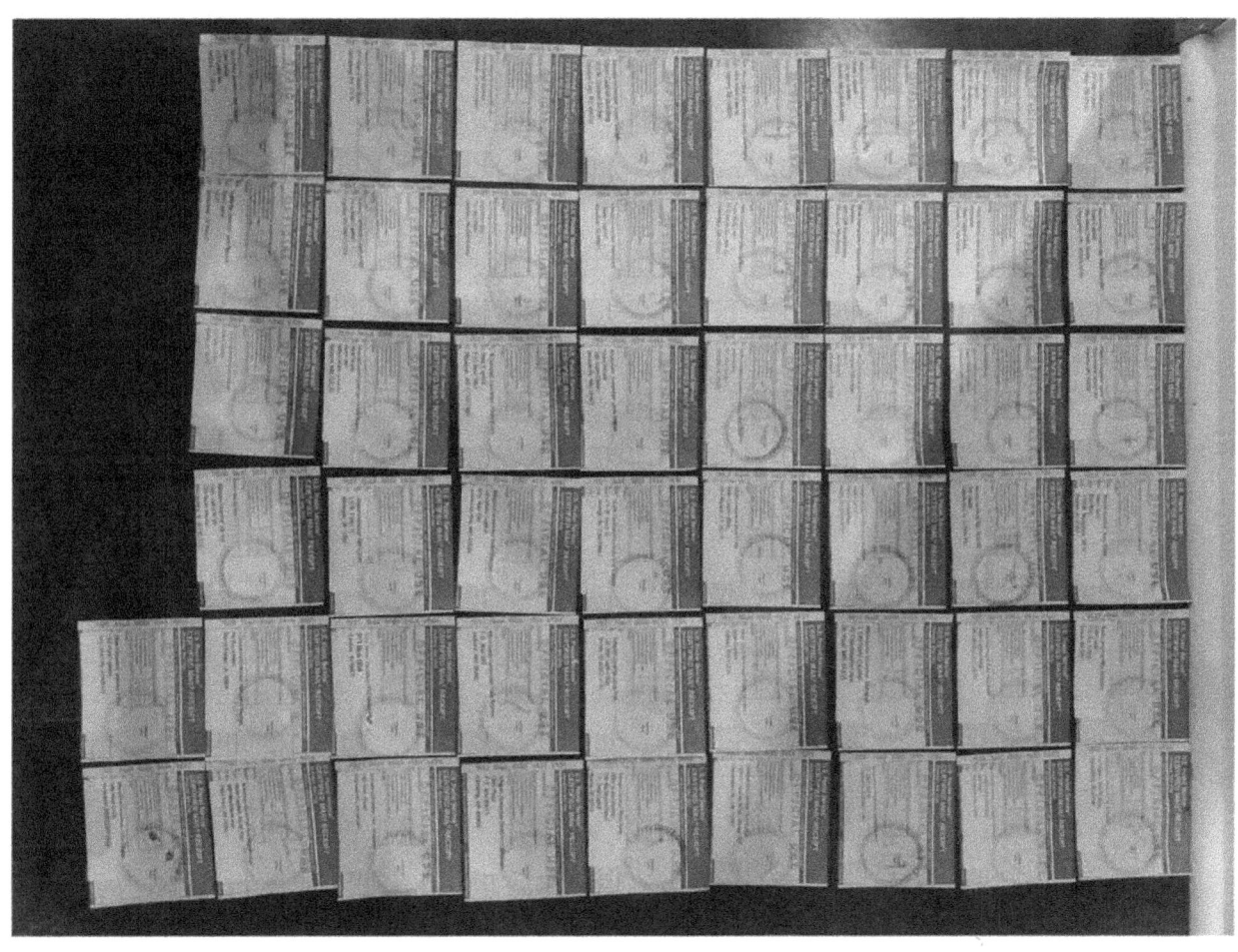

50 certified mail letters were sent addressed to the Governors (50) and their respective Constitutional County Sheriffs (3,142 all total) that informed them that the Internal Revenue Service (IRS) and British Accreditation Registry-Crown Temple British Maritain Law (B.A.R.) are un-constitutional and American Tax Payers require compensation. All 3,192 elected We The People's representatives swore an oath to God they would serve and protect the Constitution of the United States against all enemies, Foreign and Domestic. If any of the 3,192 representatives do not cooperate with the 2021 Third Continental Congress's Attorney General and remove the IRS and B.A.R. from the United States, they will be committing Perjury. Which our Lord God hates.

2021 Third Continental Congress
Office of the Secretary of State
Carpenters' Hall
Philadelphia, Pennsylvania

Greetings Honorable Alabama Governor Kay Ivey and Constitutional County Sheriffs,

The first order of business of the Third Continental Congress (TCC) is to assure each State National Citizen (SNC) of the 1776 Declaration of Independence and 1787 United States of America Constitution financial and economic retribution for the money Extorted from the SNC (a.k.a "American Tax Payers") since 1915 by the following Unconstitutional and illegal organizations and individuals (Defendants):

Violates Article I, Section 8, Clause 1 and/or Article XI of the 1787 United States of America Constitution.

1. Internal Revenue Service (IRS),
2. Bankrupted United States of America, INC. (USAINC),
3. Bankrupted United Nations, INC. (UNINC),
4. UNINC US Office of Personnel Management (OPM) (i.e. FBI, CIA, Congress, IRS, President, Supreme Court Justices, etc.)
5. British Accreditation Registry-Crown Temple British Maritain Law (B.A.R.)
6. The Central Bank of the USAINC is the Private Federal Reserve System, created by Congress in 1913.
7. Private Western Central Vatican Bank and Central Bank of London.
8. North Atlantic Treaty Organization (NATO)

All State National Citizens (a.k.a. American Tax Payers) shall not have any direct contact with any of the "Defendants" stated above. If one is approached by any individual claiming to representing the above stated "Defendants", do not be adversarial or aggressive, be polite and helpful. Report all encounters and all relevant correspondences (Postal Mail, E-Mail, Phone and personal contact reports) to the local Constitutional County Sheriffs' office.

All State National Citizens (a.k.a. American Tax Payers) shall:

1. Mail a copy your 2021 Tax Form (Personal or Business) originals to U.S. Department of the Treasury, National Payment Integrity and Resolution Center, P. O. Box 51315, Philadelphia, Pennsylvania, 19115-6314 and demand money you have paid in taxes plus fees and interest (Since 1915, Example: Ratio for every $1 Tax paid times 10 = $10/$1) to the IRS be returned to you. Do not send any money.

2. Mail a copy of your tax forms, Blackout out your Signature and Social Security Number, to one of the following three (3) private collection agencies representing the IRS.

IRS Notice CP40 and Publication 4518: Effective September 23, 2021, when the IRS assigns your account to a private collection agency, one of these three agencies will contact you on the government's behalf:

a. CBE Group Inc., P.O. Box 2217, Waterloo, IA 50704, 800-910-5837
b. Coast Professional Inc., P.O. Box 526, Albion, NY 14411, 888-928-0510
c. Conserve P.O. Box 307, Fairport, NY 14450, 844-853-4875

The Third Continental Congress Attorney General shall coordinate with the Governor of each of the 50 United States during the resolution of this Extortion Case and provide a progress report weekly.

Praise God for His gifts of Life, Liberty, and the Pursuit of Happiness to all mankind.

James N. Akins, Jr. © Without Prejudice

Once again since 07/04/1776 on 02/22/2022, the planet Pluto will be over the USA.

Is the Internal Revenue Service (IRS) and British Accreditation Registry-Crown Temple British Maritain Law (B.A.R.) Unconstitutional? It is a YES or NO answer. If your answer is NO, they are Constitutional, it is your Constitutional duty to present your undisputable proof to God and 330 million We The People Americans.

This is my proof the IRS and B.A.R Unconstitutional. So why file $1.6 trillion in taxes per year when the IRS and Courts that inforce the IRS notices are foreign agents?

To reclaim ALL YOUR money paid to the IRA since 1915, just do 2 things:

1. Every We The People tax payer mail a copy your Tax Form (Personal or Business) originals to U.S. Department of the Treasury, National Payment Integrity and Resolution Center, P. O. Box 51315, Philadelphia, Pennsylvania, 19115-6314 and demand that all the money you have paid in taxes (Since 1915, Example: Ratio for every $1 tax paid times 10 = $10/$1) to the IRS be returned to you.

2. Mail a copy of your tax forms, Blackout out your Signature and Social Security Number, to one of these 3 private collection agencies.

IRS Notice CP40 and Publication 4518:

Effective September 23, 2021, when the IRS assigns your account to a private collection agency, one of these three agencies will contact you on the government's behalf:

a. CBE Group Inc., P.O. Box 2217, Waterloo, IA 50704, 800-910-5837

b. Coast Professional Inc., P.O. Box 526, Albion, NY 14411, 888-928-0510

c. Conserve P.O. Box 307, Fairport, NY 14450, 844-853-4875

Quote the following:

IRS agents are neither trained nor paid by the Bankrupted United States of America, INC (USAINC) Government.

Pursuant to Treasury Delegation Order No. 92, the IRS is trained under the direction of the Division of Human Resources United Nations (U.N.) and the Commissioner (International), by the Office of Personnel Management.

In the 1979 edition of 22 USCA278, "The United Nations," you will find Executive Order 10422. The Office of Personnel Management (OPM) is under the direction of the Bankrupt Secretary of the United Nations, INC. (UNINC), a violation of Article I, Section 8, Clause 1 and Article XI of the 1787 United States Constitution. The OPM supports the Executive Branch - 4.25 million employees, Legislative Branch - 35,000 employees, Judicial - 34,000 employees and Uniformed Military - 1.43 million (200 million employees support the Uniformed Military). Between 1914 and 2018, We The People have been paying taxes to the Federal Reserve Bank and UNINC (1979) Treasury Departments (Not the US Treasury). The 1787 Constitutional US Treasury and US Postal Office has been moved to Pennsylvania, according to the First, Second and Third Continental Congresses. The provisional Democratic Republic of the United States of America Third Continental Congress is debt free and the US Treasury will take back the financial responsibility of US Office of Personnel Management and employees that supports the US constitutional Military.

Third Continental Congress convened on 12/22/21, in accordance with these two Articles added to the 1776 Declaration of Independences and 1787 Constitution. These two amendments were made legal by posting in a public newspaper, copyrighting in "The Story of Our Life, Based on a True Life." and entering into the Library of Congress.

Article I:
By the Grace of God - TimesDaily, Florence, Alabama - page B6, December 19, 2021
https://www.dropbox.com/s/74wfcgkmwt6wuu0/TD_MAIN_211219_B006.pdf?dl=0

Article II:
Peace at Your Gate - TimesDaily, Florence, Alabama - page A8, February 8, 2022
https://www.dropbox.com/s/ku9rf7xydse06oj/TD_MAIN_220208_A008.jpg?dl=0

A copy of "The Story of Our Life, Based on a True Life." and invitations to attend the Third Continental Congress was sent Certified mail to all 50 Governors to attend. "The Story of Our Life, Based on a True Life." contains True Quantum Physics and Math from God and technology for building a True 4-dimensional Quantum Base 2 - Base 3 Computer. Making the false Base 2- Base 10 Quantum computer obsolete. Please go to website www.piisthree.com for more information.

ebook:
https://www.dropbox.com/s/xp8vsmtwl90mwmj/1EBookstoryofourlife%2Cbasedonatruelife.pdf?dl=0

Video:
https://www.dropbox.com/s/bwta72nqozpaece/Albert%20Einstein%20Unified%20Quantum-Universe%20Laws%20of%20Physics.mp4?dl=0

A4 WEDNESDAY, MARCH 2.2022
THE ONLOOKER
GULFCOASTMEDIA.COM

Videos on Rumble.com/BubbaTwain

https://rumble.com/vburcz-albert-einsteins-dynamic-4th-dimension.html

https://rumble.com/vau1w7-albert-einstein-was-100-wrong-the-human-mind-does-concieve-of-the-four-dime.html

https://rumble.com/vatzid-the-story-of-our-life-based-on-a-true-life.-the-human-mind-concieves-the-fo.html

https://rumble.com/vatz5b-albert-einstein-and-nikola-tesla-deserve-nobel-prizes-for-four-dimensions.html

https://rumble.com/vbzg9b-the-holy-grail-of-quantum-physics-is-in-the-pyramid-of-khafus-design.html

https://rumble.com/vc0g2r-albert-einstein-and-nikola-tesla-unified-quantumuniverse-11-laws-of-physics.html

Political Advertisement

By the Grace of God

Lord Jesus Christ, on His 2021st Anniversary of His December 22nd Birthday. Our Lord has fulfilled all His promises in His New Testament with these final written words:

Zechariah 8:16-17
New American Standard Bible
16 These are the things which you shall do: speak the truth to one another; judge with truth and judgment for peace at your gates. 17 Also let none of you devise evil in your heart against another, and do not love perjury; for all these things are what I hate,' declares the Lord."

Nahum 1-3
New American Standard Bible
1. The pronouncement of Nineveh. The book of the vision of Nahum the Elkoshite:
2. A jealous and avenging God is the Lord;
The Lord is avenging and wrathful.
The Lord takes vengeance on His adversaries,
And He reserves wrath for His enemies.
3. The Lord is slow to anger and great in power,
And the Lord will by no means leave the guilty unpunished.

The following is how God will free all humanity and make them all Sovereign Souls of God around the World. This is how I know his promise is True.

NASB 1995
2 Peter 3 vs 9
The Lord is not slow about His promise, as some count slowness, but is patient toward you, not wishing for any to perish but for all to come to repentance.

Greetings to We The People of the United States of America that swore an oath to God to serve and protect the Constitution, over 200 million Sovereign Souls of God, against all enemies, foreign and domestic. Now is the time to come to the aid of your fellow People of the World. Our captors, UNITED STATES OF AMERICA CORPORATION AND UNITED NATION CORPORATION are legally bankrupted and have no more means of supporting their Tyrannical Governments.

As of September 23, 2021, the Internal Revenue Service, the tax collection agency of the Secretary of State of the bankrupted UNITED NATIONS CORPORATION has turned over their tax collection to three private collection agencies. Private tax collection agencies cannot legally garnish We The People's income and tax payers are being told to address their tax checks to the US Treasury. God is a God of Law, the UNITED STATES OF AMERICA CORPORATION has been under Martial Law since 1933. The last duly elected 45th President, during the November 3, 2020 Pearl Harbor like attack on our Presidential election, declared the beginning of World War III and was call into Service by Article II, Section 2, Clause 1 as the Commander-in-Chief of the Constitutional Military.

On December 7, 2021 at 7:48 am, all US Military personnel, officers and enlisted, under oath to the 1787 Constitution removed the gold fringe war flag from their Title 10, Subtitle A, Part II, Chapter 45 of United States Code uniform and mail it to the Secretary of State of their State of original residence. Upon placing a Patented Type 4 US Flag on their Title 10 USC uniform, all UNITED STATES OF AMERICA CORPORATION AND UNITED NATION CORPORATION combatants, foreign and domestic, shall be captured and removed from the 50 United States of America, by the Grace of God. The Honorable Post Master General of the World :RussellJay:Gould:, has demanded that All North Atlantic Treaty Organization (NATO) Military and UNITED NATIONS CORPORATION MEMBERS go to SEA and return to their Land of origin and set up a new Postal Address as a Sovereign Soul of God.

The Honorable Post Master General of the World has captured all gold trim British Crown admiralty law war flags and demand all vessels (Combatants of We The People of the World) abandon their LAND SHIP and return to the SEA. All Judges and lawyers that pledged allegiance to the BRITISH ACCREDITATION REGISTRY- CROWN TEMPLE B.A.R., which is prohibited by Article XI, have no Postal Address to do business with any We The People of the World Facilities. They all have no Postal Address and may no longer conduct CONTRACTS with We the People on God's Lands. All their LAND SHIPS (STRUCTURES and PROPERTY) shall be salvaged by We The People. All Lawyers and Judges shall return to their original Postal Address and swear oath to God to "support and defend the Constitution of the United States against all enemies, foreign and domestic."

God Save Our Souls (SOS), according to the Geneva Conventions: Combatants, engaged in military operations, have to distinguish themselves from the civilian population to protect it from the effects of hostilities and to restrict warfare to military objectives.

On December 22, 2021 and in accordance with the 1776 Declaration of Independence, all 50 Governors of the United States of America have been sent Certified Postal Mail requesting they attend the convening of the Third Continental Congress in the Carpenters' Hall of Philadelphia, Pennsylvania to re-form a more perfect Union. If the Governors do not attend, it is the right of We The People to send their State representative. Georgia did not attend the First Continental Congress. The Third Continental Congress shall provide economic stability through Common Laws and the US Treasury for all We The People Sovereign Souls of God.

All 50 Governors received a copy of The Story of Our Life, Based on A True Life. The book contains God's True Science, that will lead to a true Quantum Base 2/ Base 3 Computer and cures for all man-made illnesses. We The People of the World will declare Eminent Domain on all Corporations, Pharmaceuticals, Banks, Water, Communications and Utilities Corporations that have enslaved humanity. Most importunately, Eminent Domain shall be declared on all Freemason copyrights and patents on the Block Chain Computers, that is controlled by God until his Quantum Computers are made.

God promised to always Bless America and all Sovereign Souls of God around the World.

Please go to www.piisthree.com for more information.

By The Grace of God
We The People request the presence:
The Honorable Commander-in-Chief of the Constitutional Military
Donald J. Trump
Third Continental Congress
December 22, 2021 • 8PM

Paid political by James N. Akins, P.O. Box 1111 Fairhope, AL 36533

Political Advertisement

By the Grace of God

Lord Jesus Christ, on His 2021st Anniversary of His December 22nd Birthday. Our Lord has fulfilled all His promises in His New Testament with these final written words:

Zechariah 8:16-17
New American Standard Bible
16 These are the things which you shall do: speak the truth to one another; judge with truth and judgment for peace at your gates. 17 Also let none of you devise evil in your heart against another, and do not love perjury; for all these things are what I hate,' declares the Lord."

Nahum 1-3
New American Standard Bible
1. The pronouncement of Nineveh. The book of the vision of Nahum the Elkoshite:
2. A jealous and avenging God is the Lord;
The Lord is avenging and wrathful.
The Lord takes vengeance supporting their Tyrannical Governments.

As of September 23, 2021, the Internal Revenue Service, the tax collection agency of the Secretary of State of the bankrupted UNITED NATIONS CORPORATION has turned over their tax collection to three private collection agencies. Private tax collection agencies cannot legally garnish We The People's income and tax payers are being told to address their tax checks to the US Treasury. God is a God of Law, the UNITED STATES OF AMERICA CORPORATION has been under Martial Law since 1933. The last duly elected 45th President, during the November 3, 2020 Pearl Harbor like attack on our Presidential election, declared the beginning of World War III and was call into Service by Article II, Section 2, Clause 1 as the Commander-in-Chief of the Constitutional Military.

On December 7, 2021 at 7:48 am, all US Military 45 of United States Code uniform and mail it to the Secretary of State of their State of original residence. Upon placing a Patented Type 4 US Flag on their Title 10 USC uniform, all UNITED STATES OF AMERICA CORPORATION AND UNITED NATION CORPORATION combatants, foreign and domestic, shall be captured and removed from the 50 United States of America, by the Grace of God. The Honorable Post Master General of the World :Russell-Jay:Gould:, has demanded that All North Atlantic Treaty Organization (NATO) Military and UNITED NATIONS CORPORATION MEMBERS go to SEA and return to their Land of origin and set up a new Postal Address as a Sovereign Soul of God.

The Honorable Post Master General of the World has captured all gold trim British Crown admiralty law war flags and demand all vessels (Combatants of We The People of the World) abandon their LAND SHIP and return to the SEA. All Judges and Lawyers that pledged allegiance to the BRITISH ACCREDITATION REGISTRY- CROWN TEMPLE B.A.R., which is prohibited by Article XI, have no Postal Address to with We the People on God's Lands. All their LAND SHIPS (STRUCTURES and PROPERTY) shall be salvaged by We The People. All Lawyers and Judges shall return to their original Postal Address and swear oath to God to "support and defend the Constitution of the United States against all enemies, foreign and domestic."

God Save Our Souls (SOS), according to the Geneva Conventions: Combatants, engaged in military operations, have to distinguish themselves from the civilian population to protect it from the effects of hostilities and to restrict warfare to military objectives.

On December 22, 2021 and in accordance with the 1776 Declaration of Independence, all 50 Governors of the United States of America have been sent Certified Postal Mail requesting they attend the convening of the Third Continental Congress in the Carpenters' Hall of Philadelphia, Pennsylvania to re-form a more perfect Union. If the Governors do not attend, it is the right of We The People to send their State representative. Georgia did not attend the First Continental Congress. The Third Continental Congress shall provide economic stability through Common Laws and the US Treasury for all We The People Sovereign Souls of God.

on His adversaries,
And He reserves wrath for His enemies.
3. The Lord is slow to anger and great in power,
And the Lord will by no means leave the guilty unpunished.

The following is how God will free all humanity and make them all Sovereign Souls of God around the World. This is how I know his promise is True.

NASB 1995
2 Peter 3 vs 9
The Lord is not slow about His promise, as some count slowness, but is patient toward you, not wishing for any to perish but for all to come to repentance.

Greetings to We The People of the United States of America that swore an oath to God to serve and protect the Constitution, over 200 million Sovereign Souls of God, against all enemies, foreign and domestic. Now is the time to come to the aid of your fellow People of the World. Our captors, UNITED STATES OF AMERICA CORPORATION AND UNITED NATION CORPORATION are legally bankrupted and have no more means of personnel, officers and enlisted, under oath to the 1787 Constitution removed the gold fringe war flag from their Title 10, Subtitle A, Part II, Chapter do business with any We The People of the World Facilities. They all have no Postal Address and may no longer conduct CONTRACTS

I could think of no worse example for nations abroad, who for the first time were trying to put free electoral procedures into effect, than that of the United States wrangling over the results of our presidential election, and even suggesting that the presidency itself could be stolen by thievery at the ballot box.

Whenever any form of government becomes **destructive of these ends** [life, liberty, and the pursuit of happiness] it is the **right of the people to alter** or abolish it, and to institute new government.

Does the government fear us? Or do we fear the government? When the people fear the government, tyranny has found victory. The federal government is our servant, not our master!

The God who gave us life, gave us liberty at the same time.

Rebellion to tyranny is obedience to God.

I have sworn upon the altar of God Eternal, hostility against every form of tyranny over the mind of man.

The Bible is the cornerstone of liberty.

~Thomas Jefferson

By The Grace of God
We The People request the presence:
The Honorable Commander-in-Chief of the Constitutional Military
Donald J. Trump
Third Continental Congress
December 22, 2021 • 8PM

All 50 Governors received a copy of The Story of Our Life, Based on A True Life. The book contains God's True Science, that will lead to a true Quantum Base 2/ Base 3 Computer and cures for all man-made illnesses. We The People of the World will declare Eminent Domain on all Corporations, Pharmaceuticals, Banks, Water, Communications and Utilities Corporations that have enslaved humanity. Most importunately, Eminent Domain shall be declared on all Freemason copyrights and patents on the Block Chain Computers, that is controlled by God until his Quantum Computers are made.

God promised to always Bless America and all Sovereign Souls of God around the World.

Please go to www.piisthree.com for more information.

Paid political by James N. Akins. P.O. Box 1111 Fairhope, AL 36533

PAID POLITICAL ADVERTISEMENT

1776 Declaration of Independence and 1787 Constitution, Third Continental Congress: Article II: Peace at Your Gates.

Where We Go With =1= Christ We Go All (WWGW=1=CWGA)

Lord Jesus Christ, on His 2021st Anniversary of His December 22nd Birthday, has fulfilled all His promises in His New Testament by the convening of the Third Continental Congress Provisional Government at Carpenters' Hall Philadelphia, Pennsylvania, in accordance with the 1776 Declaration of Independence and 1787 Constitution (Article I: By the Grace of God, page B6, Sunday, December 19, 2021, TimesDaily newspaper, Florence, Alabama). To solidify the Lord's promise of 1000 years of peace at our gates, these are His final written words found in a cave near the Dead Sea late 2017:

Zechariah 8:16-17

New American Standard Bible 16 These are the things which you shall do: speak the truth to one another; judge with truth and judgment for peace at your gates. 17 Also let none of you devise evil in your heart against another, and do not love perjury; for all these things are what I hate,' declares the Lord."

Nahum 1-3

New American Standard Bible
1. The pronouncement of Nineveh. The book of the vision of Nahum the Elkoshite:
2. A jealous and avenging God is the Lord;
The Lord is avenging and wrathful.
The Lord takes vengeance on His adversaries,
And He reserves wrath for His enemies.
3. The Lord is slow to anger and great in power,
And the Lord will by no means leave the guilty unpunished.

The Lord's words shall be written into the new Constitution of the Democratic Republic of the United States of America (DR-USA), where by, no laws shall be written by the Congress of the DR-USA that will ever prevent any individual We The People from securing their Life, Liberty and Pursuit of Happiness, By the Grace of God.

The Third Continental Congress (TCC) and Provisional Government was convened on 12/22/2021. While the TCC is in session, We The People approves the following leaders of the three Branches of Government:

1. Executive Branch and 1787 Constitutional Military:

    1. 1787 Constitutional Commander- In - Chief - Donald J. Trump- call into Service by Article II, Section 2, Clause 1 on November 3, 2020.
    2. We The People Appointed Third Continental Congress President - Lt. Col. Wendy Rogers.
    3. World Postmaster General - :Russell-Jay: Gould: - Patent holder of the Type 4 United States of America Flag.
    4. TCC Sargent at Arms - Lt. General Michael Flynn - 1787 Constitutional protocol enforcer.
    5. TCC Secretary of Treasury - To be determined by the TCC President.
    6. TCC Secretary of State - James N. Akins, Jr. - Copyright holder of 'The Story of Our Life, Based on a True Story." Contains God's Base 2-Base 3 Quantum Computer technology.

2. Judicial Branch:

    1. The 9 Supreme Court Justices shall continue under the TCC if they provide written oaths to God that they will, "Protect and Defend the 1787 Constitution against all enemies, Foreign and Domestic." If a Supreme Court Judge shall vacate their position, a replacement Judge shall be appointed by the TCC President, Vetted and Approved by the all 50 State TCC Representatives.

    2. The Supreme Court shall create and implement a new Federal DR-USA Constitutional and Common Law Court System and Lawyer Registration System. The BRITISH ACCREDITATION REGISTRY- CROWN TEMPLE B.A.R., which is prohibited by Article XI, shall be removed from all the Continental United States of America and not be allowed to conduct contracts with any citizen in and of the DR-USA.

3. Legislative Branch:

    1. The 50 United States of America shall each send 2 delegates (Total 100) for Congress and 2 delegates each for The House of Representatives (Total100). Each State Legislatures, in all 50 States, shall vote on all proposed Amendments to the 1787 Constitution. It shall be determined by vote, if an existing or new amendment shall: 1. Remain the same. 2. Be amended to best comply with the 1787 Constitution and present Constitutional requirements. Or 3. Be Deleted.

    2. Any amended or new proposed Amendments to the 1787 Constitution, like Federal voting regulations, the State Legislatures shall send their recommendations to the TCC, the TCC shall compile the like suggestions, and draft the new amendments. The amendments shall be voted on by all 50 State Legislatures, as a Yes or No vote. Say a State Congress is split more than 51%, one state delegate shall vote Yes and one shall vote No. If more that 51% of the TCC delegates vote NO on an Amendment the TCC delegates can decide to re-write it or drop it.

    3. The Third Continental Congress can only be adjourned by agreement of all We The People by means of all 50 Governors ratifying the New Democratic Republic of the United States of America.

Proposed Goals of Achievement by the Third Continental Congress:

1. United States of America, INC (USAINC) and United Nations INC (UNINC) and North Atlantic Treaty Organization (NATO) Foreign and Domestic 1787 Constitution Combatants shall be removed from the land mass of the United States of America by the Commander-in-Chief Constitutional Military.
2. The TCC, like the First and Second Continental Congress, is the provisional Government until the 50 State Governors ratifies a new Democratic Republic of the United States of America.
3. The TCC owes no debts of their own and shall not accept or be legally responsible for debts of the bankrupted USAINC, UNINC or NATO.
4. To avoid paying their debtors; USAINC, UNINC, NATO, Federal Reserve Bank, Vatican City INC and City of London Central Banks, INC and others transferred funds into the US Treasury, expecting to retrieve them through their copyrights and patents on the Block Chain Computers. All funds in the US Treasury now belong to We The People of the Democratic Republic of the United of States America and will be legally returned to individual We The People citizens that it was stolen from by all the Banks through fraudulent contracts and the United Nations, INC and Central Bank's Private Tax collection agency, the Internal Revenue Service.
5. We The People shall not pay taxes to any Federal Government, including the TCC, as and until there are 1787 Constitutional requirements.

6. The existing United Nations INC, United States Office of Personnel Management (OPM) shall be terminated and transferred to TCC Secretary of State. All US Treasury funds will no longer be paid to the Bankrupted USAINC, UNINC or NATO for any reasons. All on going Contracts committed by the Bankrupted USAINC or UNINC shall be completed or terminated using only their own funds. The US Treasury shall not be legally bound to any Contract Commitments made by the Bankrupted United States of America INC or United Nations INC.

7. The TCC Office of Personnel Management shall continue paying the salaries to their employees, provide they give written oats to God that they will, "Protect and Defend the 1787 Constitution against all enemies, Foreign and Domestic." The Commander-in-Chief shall recommend to the TCC Secretary of State which persons, groups and Companies supporting the 1787 Constitutional Military and the 1787 Constitutional Military personnel, shall be paid by the US Treasury.

8. All Gold Fringe US combatant war Flags shall be removed from the Continental United States of America and replaced with a Patented Type 4 US Flag.

9. The TCC Government shall take eminent domain over all Freemason Block Chain Computers and Software. TCC Provisional Government and DR-USA shall take eminent domain over any citizen's businesses or companies, organizations, communication organizations, media organization, power or water utilities corporations, and banks that conducts their businesses in an un-constitutional manner that commits or helps to commit harm to any of the We The People citizens. Such as, but not limited to, causing communications and free speech blackouts, food or utilities shortages, and owners of deadly and health harming facilities, Hospitals, clinics and pharmaceutical manufacturers.

PAID POLITICAL BY JAMES N. AKINS, P.O. BOX 1111, FAIRHOPE AL 36533

The Secrets of The Pyramids; The Anatomy Of The Brain; The Mystery Now Revealed

This is the Height and Circumference of the earth (Kings 1 says 30 cubits Circumference and 5 cubits high. Frequency of the Earth Computer.

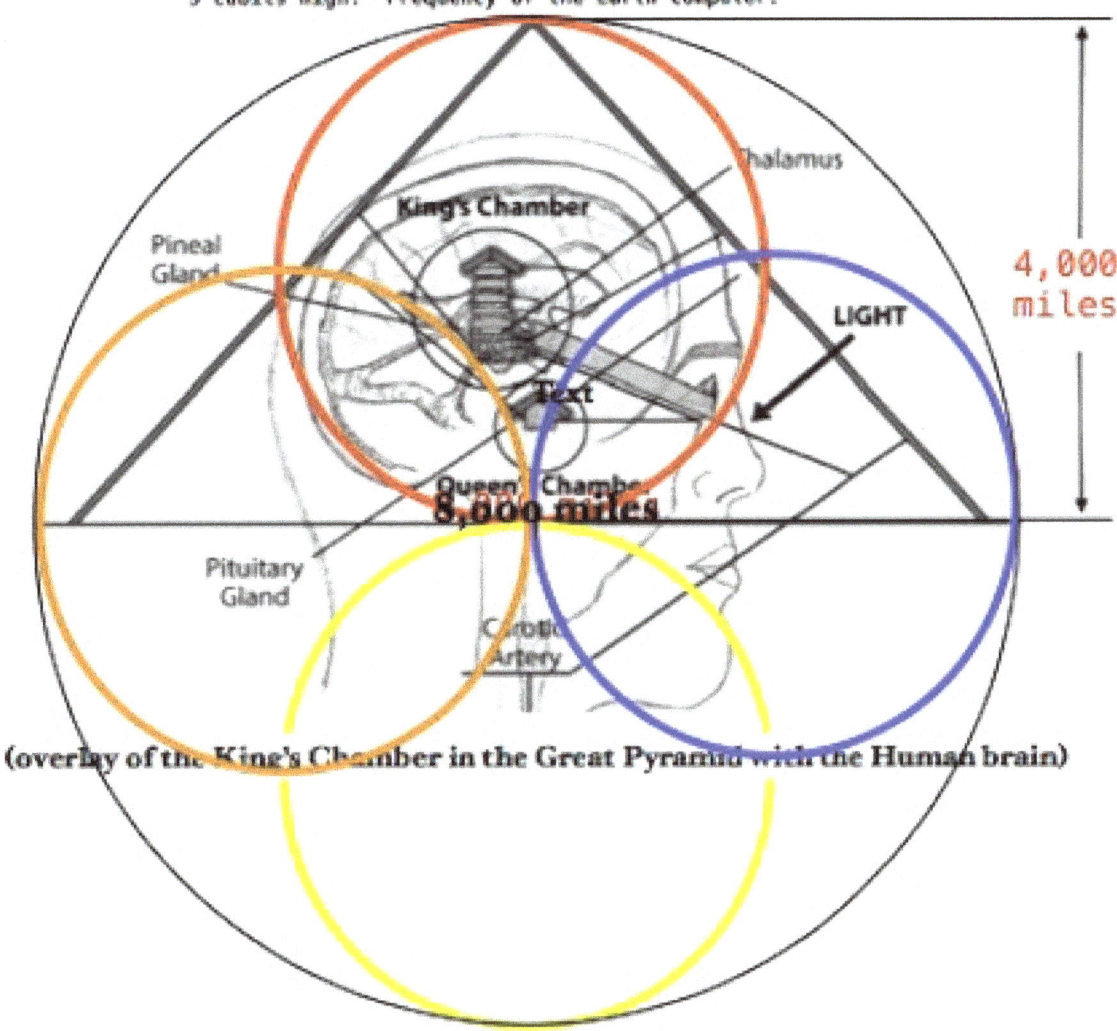

(overlay of the King's Chamber in the Great Pyramid with the Human brain)

Image taken from Alchemy Of The Gods by Michael Feeley ©

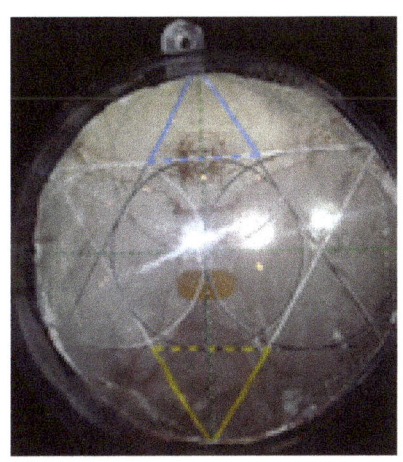
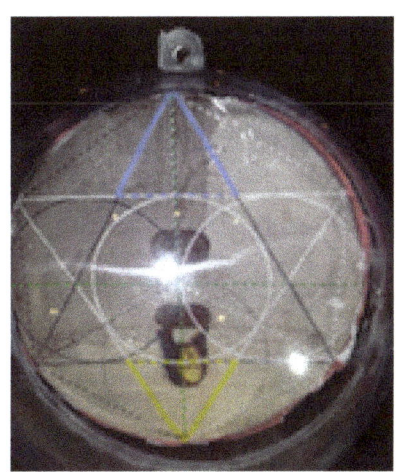

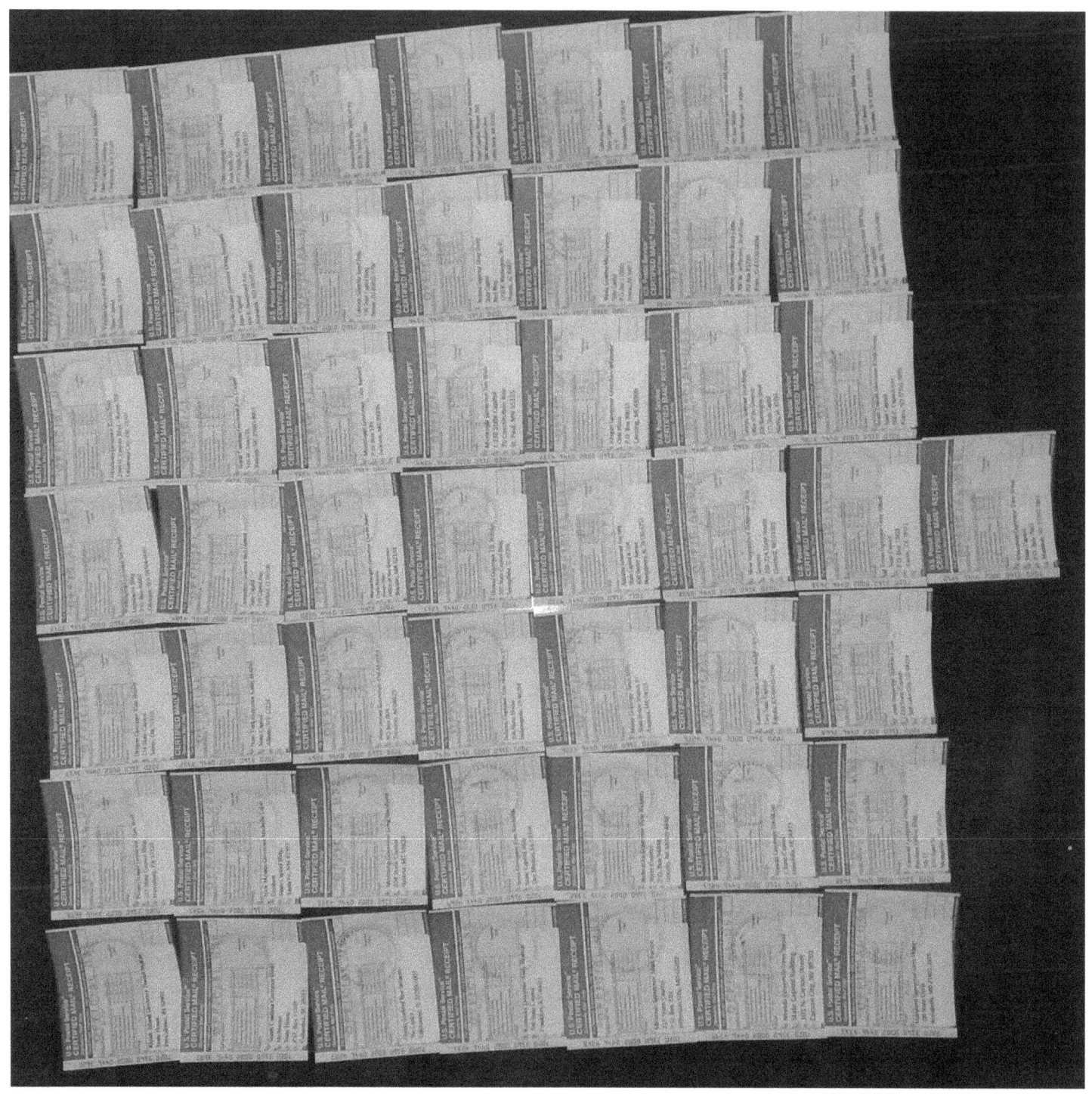

Certified mail sent to 50 United States Governors and 330 million We The People inviting everyone to the Swearing-in Ceremony of the 2021 Third Continental Congress provisional Constitutional Government on October 27, 2022 - God's Jubilee Day in Fairhope, Alabama on the fable Bay of the Holy Spirit.

2021 Third Continental Congress
Office of the Secretary of State
Carpenters' Hall Philadelphia, Pennsylvania

Greetings We The People, including 3412 Constitutional County Sheriffs, in the 50 United States of America; Alabama, Alaska, Arizona, Arkansas, California, Colorado, Connecticut, Delaware, Florida, Georgia, Hawaii, Idaho, Illinois, Indiana, Iowa, Kansas, Kentucky, Louisiana, Maine, Maryland, Massachusetts , Michigan, Minnesota, Mississippi, Missouri, Montana, Nebraska, Nevada, New Hampshire, New Jersey, New Mexico, New York, North Carolina, North Dakota, Ohio, Oklahoma, Oregon, Pennsylvania, Rhode Island, South Carolina, South Dakota, Tennessee, Texas, Utah, Vermont, Virginia, Washington, West Virginia, Wisconsin, Wyoming.

All 50 Governors did not send 4 Representatives from each State as requested by certified mail, example of request sent is on page 79, in the book titled: *The Story of Our Life, Based on a True Life.*, Library of Congress Control Number: 2022904632. Free copies of ebook in websites Piisthree.com and Piis3.com under "Our Products."

Therefore, it is We The People's God given right through the 1776 Declaration of Independence, Articles I and II and the 1787 Constitution to select 4 Representatives from each of the 50 States (Total: 200 each) and send them to the Swearing-in Ceremony in Fairhope, Alabama on the Bay of the Holy Spirit on October 27, 2022. The First Section of the Third Continental Congress shall start on December 22, 2022.

The Honorable Associate Justice Clarence Thomas shall be swearing-in:

1. 1787 Constitutional Commander-in- Chief – Honorable Donald J. Trump – called in to Service by Article 11, Section 2, Clause 1 on November 3, 2020.

2. We The People Appointed Third Continental Congress President – Honorable Lt. Col. Wendy Rogers.

3. We The People Appointed Third Continental Congress Vice President - Honorable Melissa Martz.

4. We The People Appointed Third Continental Congress Attorney General – Honorable Sidney Powell.

5. World Postmaster General - Honorable :Russell-Jay: Gould:- Patent holder of the Type 4 United States of America Flag.

6. We The People Appointed Third Continental Congress Sargent at Arms- Honorable Lt. General Michael Flynn – 1887 Constitutional protocol enforcer.

7. Third Continental Congress Secretary of the US Treasury – To be determined by the Third Continental Congress President.

8. We The People Appointed Third Continental Congress Secretary of State- James N. Akins, Jr. – Copyright holder of *The Story of Our Life, Based on a True Life.*, which contains God's Base 2 – Base 3 Dynamic 4 Dimensional Q:bit – Spherical Quantum Computer technology.

9. 200 50 State Congress Representatives.

Praise God for His gifts of Life, Liberty, and the Pursuit of Happiness to all Mankind.

James N. Akins Jr. © Without Prejudice

2021 Third Continental Congress
Carpenters' Hall
Philadelphia, Pennsylvania

Greetings to all We The People of the Great State of Wyoming;

By the Grace of God, and in accordance with the 1787 Constitution and 1776 Declaration of Independence with 2021 Articles I and II, the Third Continental Congress was convened at Carpenters' Hall in Philadelphia, Pennsylvania on December 22, 2021.

Everyone is invited to join the Second Democratic Republic of The United States of America. Your decision to join will be confirmed by your Governor and four (4) Representatives attending the Swearing-in Ceremony in Fairhope, Alabama on the Bay of the Holy Spirit.

Zechariah 8:16-17 New American Standard Bible 16 These are the things which you shall do: speak the truth to one another; judge with truth and judgment for peace at your gates. 17 Also let none of you devise evil in your heart against another, and do not love perjury; for all these things are what I hate,' declares the Lord."

Nahum 1-3 New American Standard Bible 1 The pronouncement of Nineveh. The book of the vision of Nahum the Elkoshite: 2 A jealous and avenging God is the Lord; The Lord is avenging and wrathful. The Lord takes vengeance on His adversaries, And He reserves wrath for His enemies. 3 The Lord is slow to anger and great in power, And the Lord will by no means leave the guilty unpunished.

Peter 3 vs 9 - The Lord is not slow about His promise, as some count slowness, but is patient toward you, not wishing for any to perish but for all to come to repentance.

Thomas Jefferson - I could think of no worse example for nations abroad, who for the first time were trying to put free electoral procedures into effect, than that of the United States wrangling over the results of our presidential election, and even suggesting that the presidency itself could be stolen by thievery at the ballot box. Whenever any form of government becomes destructive of these ends [life, liberty, and the pursuit of happiness] it is the right of the people to alter or abolish it, and to institute new government. Does the government fear us? Or do we fear the government? When the people fear the government, tyranny has found victory. The federal government is our servant, not our master! The God who gave us life, gave us liberty at the same time. Rebellion to tyranny is obedience to God. I have sworn upon the altar of God Eternal, hostility against every form of tyranny over the mind of man. The Bible is the cornerstone of liberty.

Praise God for His gifts of Life, Liberty, and the Pursuit of Happiness to all Mankind.

James N. Akins, Jr. © without prejudice
P. O. Box 1111
Fairhope, AL 36533-1111

9/1/22
9/2/22
RWR

RECEIVED
2/13

State of Maryland

7020 3160 0002 0446 4313

CERTIFIED MAIL

U.S. POSTAGE PAID
FCM LG ENV
SILVERHILL, AL
36576
SEP 02, 22
AMOUNT
$5.44
R2305H129630

OFFICE OF THE GOVERNOR
STATE OF ARIZONA
1700 WEST WASHINGTON STREET
PHOENIX, ARIZONA 85007

James N. Akins, Jr.
P.O. Box 1111
Fairhope, AL 36533

PRESORTED
FIRST CLASS

STATE OF ARIZONA

DOUGLAS A. DUCEY
GOVERNOR

EXECUTIVE OFFICE

October 3, 2022

Mr. James N. Akins
P.O. Box 1111
Fairhope, AL 36533

Dear Mr. Akins,

Thank you for contacting the Governor's Office.

All meeting and event requests for the Governor must be detailed and submitted online to the Scheduling Office for review and consideration.

Thank you again for contacting the Governor's Office.

Sincerely,

Governor's Office of Constituent Services
CS/jb

Sherry Sullivan, Mayor

October 5, 2022

Mr. James N. Akins, Jr.
P.O. Box 1111
Fairhope, AL 36533

Dear Mr. Akins,

I received the information you mailed my office. I wanted to let you know that I received it, thank you for sharing.

Sincerely,

Sherry Sullivan
Mayor

Fairhope City Hall • Fairhope, Alabama 36532

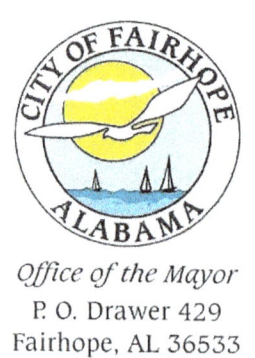

Office of the Mayor
P. O. Drawer 429
Fairhope, AL 36533

MOBILE AL 366
6 OCT 2022 PM 2 L

quadient
FIRST-CLASS MAIL
IMI
$000.57 ⁰
10/06/2022 ZIP 36532
043M31225973
US POSTAGE

Mr. James N. Akins, Jr.
P.O. Box 1111
Fairhope, AL 36533

Fort Gaines on west side mouth of Mobile Bay, Alabama.

Fort Morgan on east side mouth of Mobile Bay, Alabama.

Model of the God Particle or 4 dimensional Dynamic Quantum Computer.

I was giving a copy of my book to a young mother and her brother, hoping that they would read the last chapter and see proof that God has defeated Satan for the last time and the children of the world will no longer be harmed. I could see that they both did not believe me. So, I said I was called by Lord Jesus Christ to tell the World's People His Truths. She asked one question, "How does someone know when they have been call into service for God?"

That was the most humbling question anyone could have asked me. I do not know for sure if Our Lord Jesus Christ has called on me to teach His Blessings to His cherished Creations. I am going on pure faith and it is not my job to "sell" my words.

What I did realize is all Women were chosen by God to bring life to his Creations, including His Son Lord Jesus Christ and His Daughter Gaia.

God has blessed me beyond words. In my final version of my book I wish to record events on October 27, 2022 and December 22, 2022. My copy of my second grade class picture (1961-1962 Fairhope Alabama Elementary School) was burned in my house fire on July 5, 2016. Anyone that has a copy and knows the name of the Dark Haired Brown Eyed Girl standing next to me in the picture, please reach out to me at CONTACT under www.piisthree.com or www.piis3.com.

Also, please tell me the names of the military people and the children's pictures I took off the Internet and used in my book.

I would love to hear from the people around the world from my past life.

2021 Third Continental Congress
Carpenters' Hall
Philadelphia, Pennsylvania
September 27, 2022

Greetings to all We The People of The United States of America:

By the Grace of God, everyone is invited to join the Second Democratic Republic of The United States of America, in accordance with the 1787 Constitution and 1776 Declaration of Independence with 2021 Articles I and II, the Third Continental Congress (TCC) was convened at Carpenters' Hall in Philadelphia, Pennsylvania on December 22, 2021.

The Swearing-in Ceremony of the Third Continental Congress Executive Branch will be in Fairhope, Alabama on the Bay of the Holy Spirit on October 27, 2022. Since none of the 50 State Governors informed the 330 million The United States of America State National Citizens (SNC) about the convening of the TCC, the new 50 Governors and 200 Third Continental Congress State Representatives shall be sworn-in on December 22, 2022 at the start of the 1st Session of Congress.

The 2022 Fiscal year ends September 30, 2022 and the United Nations, Inc. will stop paying the US Office of Personnel Management's employees' and retiree's wages and Bankrupted United States of America, Inc. debts. Therefore, on October 1, 2022, The Bankrupted United States of America, Inc. will no longer exist and the Third Continental Congress provisional government (Executive Branch sworn upon the altar of God Eternal, to protect and defend the Constitution against all enemies, foreign and domestic; TCC President Wendy Rogers, Commander-in-Chief of the Continental Military President Donald J. Trump, TCC Attorney General Sidney Powell, TCC Sargent at Arms Lt. General Michael Flynn, and TCC Secretary of State James N. Akins, Jr.) shall provide for the security and economic stability of We The People of The United States of America until the Second Democratic Republic of The United States of America is ratified by all 50 States.

The Feast of the Jubilee is biblical. As one can see on the chart, Chirst's Reign Begins on the 50th year of the 120th Jubilee (precisely on December 22, 2022).

In the Bay of the Holy Spirit (Mobile Bay), we have always had Jubilees (also in Tokyo Bay).

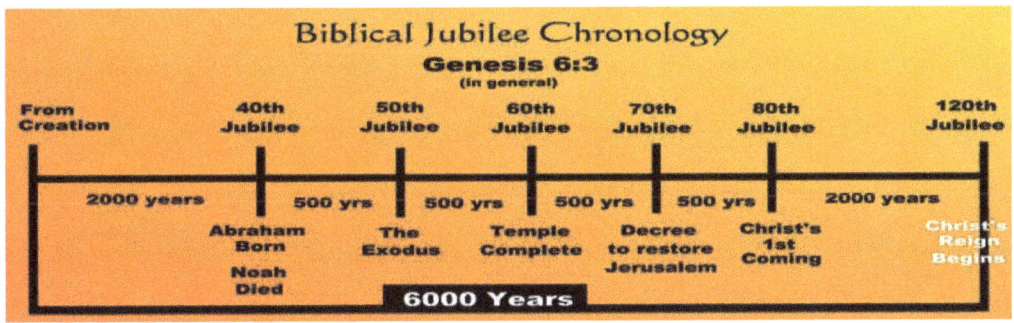

Praise God for His gifts of Life, Liberty, and the Pursuit of Happiness to all Mankind.

God is truly a loving Father. He said He would bring all the Evil to an end in 1 day and it will be like Evil never existed. The Holy Bible is not a story book it is a history book. The Holy Bible says no one knows the hour and day Christ will return, but Genesis 6:3, Biblical Jubilee Chronology, specifically states Christ's Reign Begins on the 120th Jubilee on Christ's 2022th anniversary 12 AM Jerusalem time on December 22, 2022 end of Winter Solstice.

The Fake Block Chain Quantum Computer is made up of Blocks of Computers in a chain that goes from center 0 and sends DATA from 0 to negative forever and positive forever and back, going through each Block and back to zero. The DATA is decoded by a seperate Binary- Base 10 Computer.

Biblical Jubilee Chronology is the time line of God's Quantum Computer. Picture looking at a circle at the end with a light that is moving clockwise, lighting at every Degrees 0 to 360, for infinity. Although the light is moving around the circle it is not going anywhere. Then after the light reaches a circular speed of the Constant Speed of Light and the time between each Degree is 1 DAY. Right after the light hits 360 Degrees (360 Days) the light breaks loose moving forward at the Constant Speed of Light moving in a spiral. The First Day 360 Degrees in 4 Year the Circle gets bigger, Day 2 in 4 Years the circle gets bigger on to 6 Days 24 years at 366 Days at 360 Degrees the light spirals ¼ a Day in 1 Year smaller until 360 Degrees is 365 Days then the light spirals ¼ a Day for 4 Years back to 366 Days. 1 Jubilee is 49 Years + 1 Year. Jubilee always starts at 12 AM Jerusalem on December 22 first day after Winter Solstice. It is a Jubilee because God's Quantum Computer deletes all lies, slavery and replaces what is stolen. I short the Quantum Computer resets to 0 negative Frequencies.

A Block Chain Quantum Computer and Binary- Base 10 Computer Compounds the error like the interest on money.

I know it is the 120th Jubilee on December 22, 2022, because United States of America Cooperation started Between 1871 and !872 – 117th Jubilee, and Nixon took the USA. INC off the Gold standard between 1971 and 1972- 119th Jubilee. Both of these events is the negative to a Jubilee, compounding the slavery of the world.

God freed His children from Egypt at the Red Sea, so how can we doubt He will not free His children again with a Crimson (Red) Tide this time. Only this time, if a person repents before December 22, 2022 they will be saved too.

From 0 to 3000 years it has increased 11 years, Divide 11 years by 3000 years one gets 366. Then between 3011 and 6022 Years one finds another 11 Years. God's Quantum Computer self-corrects every Jubilee. Also, if one creates a matrix of 7 cubes x 7 cubes = 49 cubes then put 1 cube over the center cube one has -1 cube and + 1 cube or 11. If the cubes are God Particles one will see one center sphere (-1+1) with 3 3 concentric circles around it. This is represented by a 7 candle menorah or negative Infinity Counter clockwise rotating Base 2 side of God's Quantum Computer. 9 cubes x 9 cubes = 81 cubes. The 1 center cube is split into -1 (1/2 cube) +1 (1/2 cube) represented by the 9 candle menorah or positive infinity Clockwise rotating Base 3 side if God's Quantum Computer,

God Bless Everyone that blesses His children.

2021 Third Continental Congress
Office of the Secretary of State
Carpenters' Hall
Philadelphia, Pennsylvania
December 19, 2022

Supreme Court of the United States
1 First Street, NE
Washington, D. C. 20543

ATTN: Honorable Nine (9) Supreme Court Justices

I support Brunson v. Alma S. Adams, et al, Case No. 22-380 with following THE documentation:

In 1871, Congress did expressly incorporate the District of Columbia, but D.C. and the "United States" are not one and the same. In that Act of 1871, Congress also expressly extended the U.S. Constitution into D.C.: Everett C. Gilbertson v. United States of America, Case No. 97-2099-MNST. Therefore, all elected and paid State and Federal Government employees are bound to their oath to "protect and defend the U.S. Constitution against all enemies, Foreign and Domestic."

The book, "The Story of Our Life, Based on a True Life." : recorded in the Library of Congress Control Number: 2022904632, Dec. 2022 printing, chronicles God's fulfillment of His victory over Satan, foretold in Genesis 6:3 - 120th Jubilee and Daniel 2:35 - Beginning of the 3rd day (1000 years of Pease).

All 50 Governors were informed in writing, through the U.S Postal Service Certified Mail, The Internal Revenue Service (IRS) and British Accreditation Registry – Crown Temple (B.A.R.) are unconstitutional, Article I, Section 8, Clause 1 and Article XI, respectively, of the U. S. Constitution. Between 2018 and October 1, 2022, the United Nations has been paying the salaries of the U. S. Office of Personnel Management employees, retirees and debts of the bankrupted United States of America, INC, the same as the elected and paid Congress and Executive Branch employees in Brunson v. Alma S. Adams, et al, Case No. 22-380.

As of Midnight December 21, 2022, bankrupted United States of America, INC. will expire, since it no longer has a source of income. All 50 Governors, violated their oath to " protect and defend the U. S. Constitution against all enemies, Foreign and Domestic.", since they failed to inform We the People of unconstitutional laws under the U. S. Constitution.

On December 22, 2021, We The People convened The Third Continental Congress, to provide security and economic stability through the U. S. Constitution and the U.S. Treasury for all We The People Sovereign Souls of God. According to 1776 Declaration of Independence and 1787 United States of America Constitution, Article II: Peace at Your Gates, February 8, 2022, no more than 9 standing Supreme Court Justices, up holding their oath to God to protect and defend the 1787 U. S. Constitution against all enemies, Foreign and Domestic, shall continue protecting and defending the 1787 U. S. Constriction as the Judicial Branch of the 2021 Third Continental Congress during the forming of the Second Constitutional Republic of The United States of America.

Anyone can go to www.piisthree.com and www.piis3.com , scroll down to "Our Products" and click on FREE ebook and FREE Video, download Free copies.

Link to IRS money claim.
https://www.dropbox.com/s/ra7ti3c369ngj8a/f1040s.pdf?dl=0

Link to FREE ebook:
https://www.dropbox.com/s/ft2mlu6y2yjm6uj/TheStoryOfOurLife.BasedOnATrueLife-eBook.pdf?dl=0

Link to FREE video:
https://www.dropbox.com/s/bwta72nqozpaece/Albert%20Einstein%20Unified%20Quantum-Universe%20Laws%20of%20Physics.mp4?dl=0

James N. Akins, Jr. © Without Prejudice
P. O. BOX 1111
Fairhope, Alabama 36533-1111
251-300-1624
Mr.akins21@yahoo.com

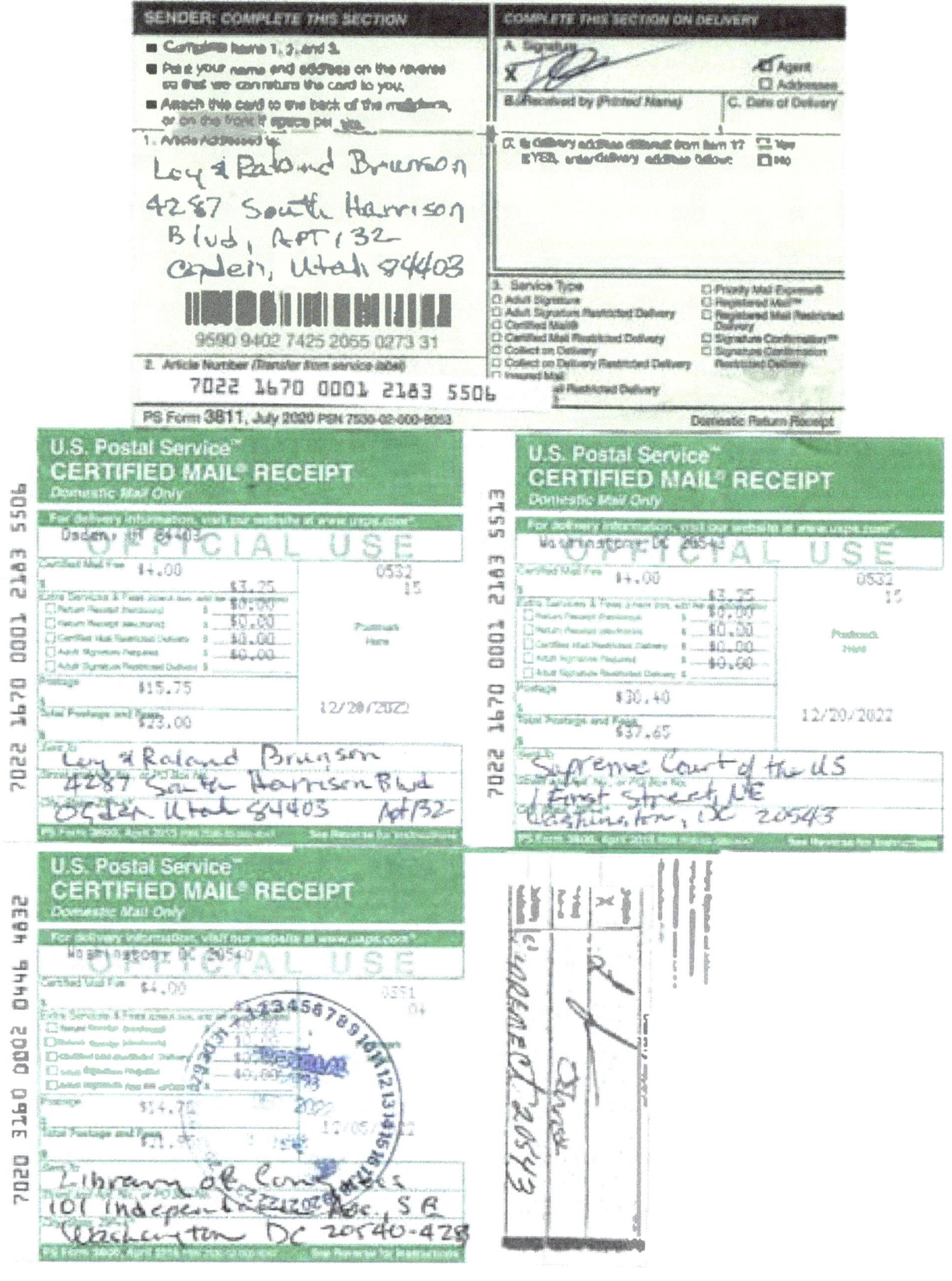

2021 Third Continental Congress
Office of the Secretary of State
Carpenters' Hall
Philadelphia, Pennsylvania
January 13, 2023

Supreme Court of the United States
1 First Street, NE
Washington, D. C. 20543

ATTN: Honorable Chief Justice John G. Roberts, Jr
       Honorable Associate Justice Sonia Sotomayor
       Honorable Associate Justice Clarence Thomas
       Honorable Associate Justice Samuel A. Alito, Jr.
       Honorable Associate Justice Elena Kagan
       Honorable Associate Justice Amy Coney Barrett
       Honorable Associate Justice Neil M. Gorsuch
       Honorable Associate Justice Brett M. Kavanaugh
       Honorable Associate Justice Ketanji Brown Jackson.

I support Brunson v. Alma S. Adams, et al, Case No. 22-380 with the following documentation:

In 1871, Congress did expressly incorporate the District of Columbia, but D.C. and the "United States" are not one and the same. In that Act of 1871, Congress also expressly extended the U.S. Constitution into D.C.: Everett C. Gilbertson v. United States of America, Case No. 97-2099-MNST. Therefore, all elected and paid State and Federal Government employees are bound to their oath to "protect and defend the U.S. Constitution against all enemies, Foreign and Domestic."

The book, "The Story of Our Life, Based on a True Life." : recorded in the Library of Congress Control Number: 2022904632, Dec. 2022 printing, chronicles God's fulfillment of His victory over Satan, foretold in Genesis 6:3 - 120th Jubilee and Daniel 2:35 - Beginning of the 3rd day (1000 years of Pease).

All 50 Governors were informed in writing, through the U.S Postal Service Certified Mail, The Internal Revenue Service (IRS) and British Accreditation Registry – Crown Temple (B.A.R.) are unconstitutional, Article I, Section 8, Clause 1 and Article XI, respectively, of the U. S. Constitution. Between 2018 and October 1, 2022, the United Nations has been paying the salaries of the U. S. Office of Personnel Management employees, retirees and debts of the bankrupted United States of America, INC, the same as the elected and paid Congress and Executive Branch employees in Brunson v. Alma S. Adams, et al, Case No. 22-380.

As of Midnight December 21, 2022, bankrupted United States of America, INC. and District of Colombia INC expire, since it no longer has a source of income. All 50 Governors, violated their oath to "protect and defend the U. S. Constitution against all enemies, Foreign and Domestic.", since they failed to inform We the People of The United States of unconstitutional laws under the U. S. Constitution.

On December 22, 2021, We The People of The United States of America convened 2021 The Third Continental Congress, to provide security and economic stability through the U. S. Constitution and the U.S. Treasury for all We The People of The United States of America's Sovereign Souls of God. According to 1776 Declaration of Independence and 1787 United States of America Constitution, Article II: Peace at Your Gates, February 8, 2022, no more than 9 standing Supreme Court Justices, up holding their oath to God to protect and defend the 1787 U. S. Constitution against all enemies, Foreign and Domestic, shall continue protecting and defending the 1787 U. S. Constriction, as the Judicial Branch of the 2021

Third Continental Congress , during the forming of the Second Constitutional Republic of The United States of America.

I respectfully request the Honorable 9 standing Supreme Court Judges render a decision on the Brunson v. Alma S. Adams, et al, Case No. 22-380 and legally and respectively terminate the bankrupt United States of America, INC and bankrupt 1871 District of Columbia, INC. By the Grace of God, beginning at 12 Noon Central Standard Time on March 4, 2023, the First Session of the 2021 Third Continental Congress shall be sworn-in to assure the security and economic stability of We the People of The United States of America, as defined by the 1876 US Constitution, until the Second Constitutional Republic of The United States of America is ratified by all 50 States.

Anyone can go to www.piisthree.com and www.piis3.com , scroll down to "Our Products" and click on FREE ebook and FREE Video, download Free copies.

Link to IRS money claim.
https://www.dropbox.com/s/ra7ti3c369ngj8a/f1040s.pdf?dl=0

Link to FREE ebook:
https://www.dropbox.com/s/ft2mlu6y2yjm6uj/TheStoryOfOurLife.BasedOnATrueLife-eBook.pdf?dl=0

Link to FREE video:
https://www.dropbox.com/s/bwta72nqozpaece/Albert%20Einstein%20Unified%20Quantum-Universe%20Laws%20of%20Physics.mp4?dl=0

James N. Akins, Jr. © Without Prejudice
P. O. BOX 1111
Fairhope, Alabama 36533-1111
251-300-1624
Mr.akins21@yahoo.com

2021 Third Continental Congress
Carpenters' Hall
Philadelphia, Pennsylvania
January 19, 2023

Greetings to all We The People of The United States of America:

By the Grace of God, all We the People of The United States of America are hereby invited to join the Second Constitutional Republic of The United States of America, in accordance with the 1787 US Constitution and 1776 Declaration of Independence with 2021 Article I and 2022 Article II. The Third Continental Congress (TCC) was convened at Carpenters' Hall in Philadelphia, Pennsylvania on December 22, 2021.

The 2021 Third Continental Congress accepts the membership of the Honorable Arkansas Governor Sarah Huckabee Sanders, and four (4) vetted TCC State Congress representatives. The names of the 4 TCC State Congress representatives shall be submitted NLT February 9, 2023 for vetting. To the best of my knowledge, no TCC accepted member is named in the Supreme Court Case No. 22-380: Brunson v. Alma S. Adams, et al.

On March 4th, 2023, the Swearing-in Ceremony for the 1st Session of the 2021 Third Continental Congress' 50 Governors and 200 Third Continental Congress State Representatives is scheduled to be in Fairhope, Alabama, on the Bay of the Holy Spirit.

The Feast of the Jubilee is biblical. As one can see on the chart, Christ's Reign Begins on the last day of the 50th year of the 120th Jubilee (precisely on December 22, 2022).

In the Bay of the Holy Spirit (Mobile Bay), we have always had Jubilees (also in Tokyo Bay).

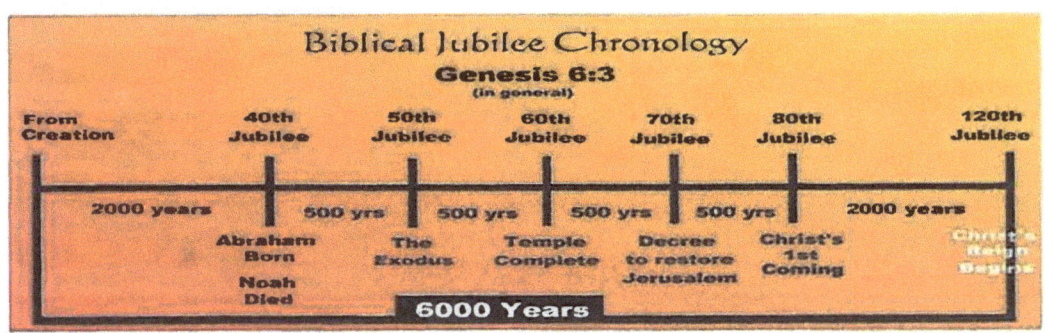

Praise God for His gifts of Life, Liberty, and the Pursuit of Happiness to all Mankind.

James N. Akins, Jr. © **Without Prejudice**
P. O. BOX 1111
Fairhope, Alabama 36533-1111
251-300-1624
Mr.akins21@yahoo.com

Form **1040-SR** Department of the Treasury—Internal Revenue Service (99) **2021** OMB No. 1545-0074 IRS Use Only—Do not write or staple in this space.
U.S. Tax Return for Seniors

Filing Status
Check only one box.

☑ Single ☐ Married filing jointly ☐ Married filing separately (MFS)
☐ Head of household (HOH) ☐ Qualifying widow(er) (QW)

If you checked the MFS box, enter the name of your spouse. If you checked the HOH or QW box, enter the child's name if the qualifying person is a child but not your dependent ▶

Your first name and middle initial	Last name	Your social security number
If joint return, spouse's first name and middle initial	Last name	Spouse's social security number

Home address (number and street). If you have a P.O. box, see instructions. | Apt. no.

City, town, or post office. If you have a foreign address, also complete spaces below. | State | ZIP code

Foreign country name | Foreign province/state/county | Foreign postal code

Presidential Election Campaign
Check here if you, or your spouse if filing jointly, want $3 to go to this fund. Checking a box below will not change your tax or refund. ☐ You ☐ Spouse

At any time during 2021, did you receive, sell, exchange, or otherwise dispose of any financial interest in any virtual currency? . . . . ▶ ☐ Yes ☐ No

Standard Deduction

Someone can claim: ☐ You as a dependent ☐ Your spouse as a dependent
☐ Spouse itemizes on a separate return or you were a dual-status alien

Age/Blindness {
  You: ☐ Were born before January 2, 1957 ☐ Are blind
  Spouse: ☐ Was born before January 2, 1957 ☐ Is blind
}

Dependents (see instructions):

(1) First name Last name	(2) Social security number	(3) Relationship to you	(4) ✔ if qualifies for (see instructions):		
			Child tax credit	Credit for other dependents
			☐	☐
			☐	☐
			☐	☐
			☐	☐

If more than four dependents, see instructions and check here ▶ ☐

Attach Schedule B if required.

1	Wages, salaries, tips, etc. Attach Form(s) W-2			**1**	
2a	Tax-exempt interest .	**2a**	**b** Taxable interest . .	**2b**	
3a	Qualified dividends . .	**3a**	**b** Ordinary dividends .	**3b**	
4a	IRA distributions . . .	**4a**	**b** Taxable amount . .	**4b**	
5a	Pensions and annuities	**5a**	**b** Taxable amount . .	**5b**	
6a	Social security benefits .	**6a**	**b** Taxable amount . .	**6b**	
7	Capital gain or (loss). Attach Schedule D if required. If not required, check here . . . . ▶ ☐			**7**	
8	Other income from Schedule 1, line 10			**8**	
9	Add lines 1, 2b, 3b, 4b, 5b, 6b, 7, and 8. This is your **total income** . . ▶			**9**	
10	Adjustments to income from Schedule 1, line 26			**10**	
11	Subtract line 10 from line 9. This is your **adjusted gross income** . . ▶			**11**	

For Disclosure, Privacy Act, and Paperwork Reduction Act Notice, see separate instructions. Cat. No. 71930F Form **1040-SR** (2021)

Form 1040-SR (2021) Page **2**

Standard Deduction See *Standard Deduction Chart* on the last page of this form.	**12a**	**Standard deduction or itemized deductions** (from Schedule A)	12a	
	b	Charitable contributions if you take the standard deduction (see instructions)	12b	
	c	Add lines 12a and 12b	12c	
	13	Qualified business income deduction from Form 8995 or Form 8995-A .	13	
	14	Add lines 12c and 13	14	
	15	**Taxable income.** Subtract line 14 from line 11. If zero or less, enter -0- .	15	
	16	**Tax** (see instructions). Check if any from: **1** ☐ Form(s) 8814 **2** ☐ Form 4972 **3** ☐ _____	16	
	17	Amount from Schedule 2, line 3	17	
	18	Add lines 16 and 17	18	
	19	Nonrefundable child tax credit or credit for other dependents from Schedule 8812	19	
	20	Amount from Schedule 3, line 8	20	
	21	Add lines 19 and 20	21	
	22	Subtract line 21 from line 18. If zero or less, enter -0-	22	
	23	Other taxes, including self-employment tax, from Schedule 2, line 21 . .	23	
	24	Add lines 22 and 23. This is your **total tax** . . . . ▶	24	
	25	Federal income tax withheld from:		
	a	Form(s) W-2	25a	
	b	Form(s) 1099	25b	
	c	Other forms (see instructions)	25c	
	d	Add lines 25a through 25c	25d	
	26	**Estimated Tax Payments between 1915 through 2022.** .	26	750,000.00
If you have a qualifying child, attach Sch. EIC.	**27a**	Earned income credit (EIC)	27a	
		Check here if you were born after January 1, 1998, and before January 2, 2004, and you satisfy all the other requirements for taxpayers who are at least age 18 to claim the EIC. See instructions . . . ▶ ☐		
	b	Nontaxable combat pay election .	27b	
	c	Prior year (2019) earned income .	27c	
	28	Refundable child tax credit or additional child tax credit from Schedule 8812	28	
	29	American opportunity credit from Form 8863, line 8 .	29	
	30	Recovery rebate credit. See instructions	30	
	31	Amount from Schedule 3, line 15	31	
	32	**Fees and Interest – Line 26 times 10** **and refundable credits** . . . . ▶	32	7,500,000.00
	33	Add lines 25d, 26, and 32. These are your **total payments** . . . . . ▶	33	8,250,000.00

Go to *www.irs.gov/Form1040SR* for instructions and the latest information. Form **1040-SR** (2021)

Form 1040-SR (2021) — Page 3

Refund	34	If line 33 is more than line 24, subtract line 24 from line 33. This is the amount you **overpaid**	34	8.250,000.00
	35a	Amount of line 34 you want **refunded to you.** If Form 8888 is attached, check here . . . . ▶ ☐	35a	8,250,000.00

Direct deposit? See instructions.
▶ b Routing number [ ][ ][ ][ ][ ][ ][ ][ ][ ] ▶ c Type: ☐ Checking ☐ Savings
▶ d Account number [ ][ ][ ][ ][ ][ ][ ][ ][ ][ ][ ][ ][ ][ ][ ][ ][ ]

	36	Amount of line 34 you want **applied to your 2022 estimated tax** . . . . ▶	36	
Amount You Owe	37	**Amount you owe.** Subtract line 33 from line 24. For details on how to pay, see instructions . . . . ▶	37	
	38	Estimated tax penalty (see instructions) . . . . . ▶	38	

Third Party Designee Do you want to allow another person to discuss this return with the IRS? See instructions . . . . ▶ ☐ **Yes.** Complete below. ☐ **No**

Designee's name ▶   Phone no. ▶   Personal identification number (PIN) ▶ [ ][ ][ ][ ][ ]

Sign Here

Under penalties of perjury, I declare that I have examined this return and accompanying schedules and statements, and to the best of my knowledge and belief, they are true, correct, and complete. Declaration of preparer (other than taxpayer) is based on all information of which preparer has any knowledge.

Joint return? See instructions. Keep a copy for your records.

Your signature	Date	Your occupation **Retired**	If the IRS sent you an Identity Protection PIN, enter it here (see inst.) [ ][ ][ ][ ][ ][ ]
Spouse's signature. If a joint return, **both** must sign.	Date	Spouse's occupation	If the IRS sent your spouse an Identity Protection PIN, enter it here (see inst.) [ ][ ][ ][ ][ ][ ]
Phone no.		Email address	

Paid Preparer Use Only

Preparer's name	Preparer's signature	Date	PTIN	Check if: ☐ Self-employed
Firm's name ▶				Phone no.
Firm's address ▶				Firm's EIN ▶

Go to *www.irs.gov/Form1040SR* for instructions and the latest information.   Form **1040-SR** (2021)

Send original to: U.S. Department of the Treasury
National Payment Integrity and Resolution Center
P. O. Box 51315
Philadelphia, Pennsylvania 19115-6314

Print copy WITHOUT Social Security and Signature and send copy to:

CBE Group Inc.
P.O. Box 2217
Waterloo, IA 50704

OR

Coast Professional Inc.
P. O. Box 526
Albion, NY 14411

OR

Conserve
P. O. Box 307
Fairport, NY 14450

2021 Third Continental Congress
Carpenters' Hall
Philadelphia, Pennsylvania
January 19, 2023

Greetings to all We The People of The United States of America:

By the Grace of God, all We the People of The United States of America are hereby invited to join the Second Constitutional Republic of The United States of America, in accordance with the 1787 US Constitution and 1776 Declaration of Independence with 2021 Article I and 2022 Article II. The Third Continental Congress (TCC) was convened at Carpenters' Hall in Philadelphia, Pennsylvania on December 22, 2021.

The 2021 Third Continental Congress extends to the Honorable John C. White, Jr., from the Great State of Arkansas, the position of 2021 Third Continental Congress House of Representatives: Speaker of the House. You shall submit your position acceptance NLT February 9, 2023 for vetting.

On March 4th, 2023, the Swearing-in Ceremony for the 1st Session of the 2021 Third Continental Congress is scheduled to be in Fairhope, Alabama, on the Bay of the Holy Spirit.

The Feast of the Jubilee is biblical. As one can see on the chart, Christ's Reign Begins on the last day of the 50th year of the 120th Jubilee (precisely on December 22, 2022).

In the Bay of the Holy Spirit (Mobile Bay), we have always had Jubilees (also in Tokyo Bay).

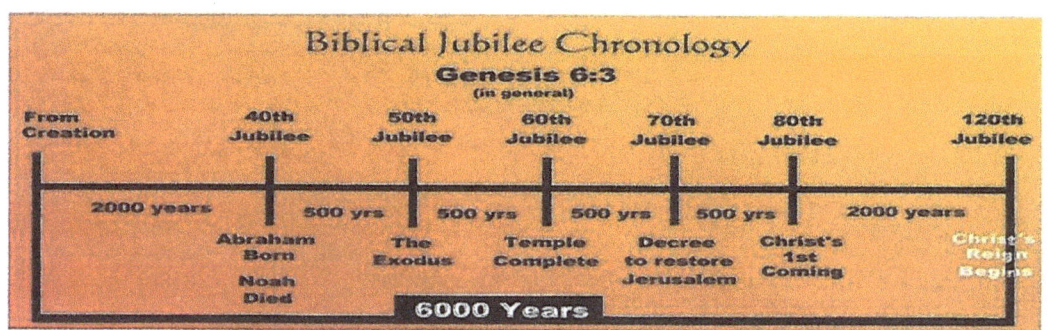

Praise God for His gifts of Life, Liberty, and the Pursuit of Happiness to all Mankind.

James N. Akins, Jr. © Without Prejudice
P. O. BOX 1111
Fairhope, Alabama 36533-1111
251-300-1624
Mr.akins21@yahoo.com

2021 Third Continental Congress
Office of the Secretary of State
Carpenters' Hall
Philadelphia, Pennsylvania
January 20, 2023

Greetings U. S. Department of the Treasury,

After March 4, 2023, you will be contacted by the Third Continental Congress (TCC) Attorney General for further instruction. The Third Continental Congress' first order of business is to assure each State National Citizen (SNC) of the 1776 Declaration of Independence and 1787 United States of America Constitution financial and economic retribution for the money Extorted from the SNC (a.k.a "American Tax Payers") since 1915 by the following Unconstitutional and illegal organizations and individuals (Defendants):

Violates Article I, Section 8, Clause 1 and/or Article XI of the 1787 United States of America Constitution.

1. Internal Revenue Service (IRS),
2. Bankrupted United States of America, INC. (USAINC),
3. Bankrupted United Nations, INC. (UNINC),
4. UNINC US Office of Personnel Management (OPM) (i.e. FBI, CIA, Congress, IRS, President, Supreme Court Justices, etc.)
5. British Accreditation Registry-Crown Temple British Maritain Law (B.A.R.)
6. The Central Bank of the USAINC is the Private Federal Reserve System, created by Congress in 1913.
7. Private Western Central Vatican Bank and Central Bank of London.
8. North Atlantic Treaty Organization (NATO)

All State National Citizens (a.k.a. American Tax Payers) shall not have any direct contact with any of the "Defendants" stated above. If one is approached by any individual claiming to representing the above stated "Defendants", do not be adversarial or aggressive, be polite and helpful. Report all encounters and all relevant correspondences (Postal Mail, E-Mail, Phone and personal contact reports) to the local Constitutional County Sheriffs' office. National Payment Integrity and Resolution Center shall retain all correspondence from both the American Tax payers and Defendants until requested by the TCC Attorney General.

All State National Citizens (a.k.a. American Tax Payers) shall:

1. Mail a copy your 2021 Tax Form (Personal or Business) originals to U.S. Department of the Treasury, National Payment Integrity and Resolution Center, P. O. Box 51315, Philadelphia, Pennsylvania, 19115-6314 and demand that all the money you have paid in taxes plus fees and interest (Since 1915, Example: Ratio for every $1 Tax paid times 10 = $10/$1) to the IRS be returned to you. Do not send any money.

2. Mail a copy of your tax forms, Blackout out your Signature and Social Security Number, to one of the following three (3) private collection agencies representing the IRS.

IRS Notice CP40 and Publication 4518: Effective September 23, 2021, when the IRS assigns your account to a private collection agency, one of these three agencies will contact you on the government's behalf:

a. CBE Group Inc., P.O. Box 2217, Waterloo, IA 50704, 800-910-5837
b. Coast Professional Inc., P.O. Box 526, Albion, NY 14411, 888-928-0510
c. Conserve, P.O. Box 307, Fairport, NY 14450, 844-853-4875

The Third Continental Congress Attorney General shall coordinate with the Governor of each of the 50 United States during the resolution of this Extortion Case and provide a progress report weekly.

All counter proposals from the Defendants, like Exhibit 1, shall be sent directly to: U. S. Department of the Treasury; National Payment Integrity and Resolution Center; P. O. Box 51315; Philadelphia, Pennsylvania 19115-6314.

Praise God for His gifts of Life, Liberty, and the Pursuit of Happiness to all mankind.

James N. Akins, Jr.
James N. Akins, Jr. © Without Prejudice, P. O. Box 1111, Fairhope, Alabama 36533-1111

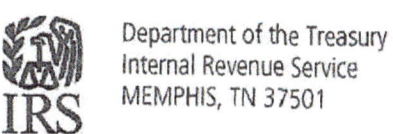

Department of the Treasury
Internal Revenue Service
MEMPHIS, TN 37501

331556.490049.307224.6812 1 AV 0.455 372

JAMES N AKINS JR
PO BOX 1111
FAIRHOPE AL 36533-1111

331556

Statement Showing Interest Income from the Internal Revenue Service	Calendar Year
(Please keep this copy for your records)	2022
Recipient's Identification Number XXX-XX-5199	Total Interest Paid or Credited $168.12
PAYER'S Federal Identification Number 38-1798424 (INTERNAL REVENUE USE ONLY)	

Form 1099-INT (Rev. 10-2013)

THIS IS NOT A TAX BILL. It shows the taxable interest paid to you during the calendar year by the Internal Revenue Service. If you are required to file a tax return, report this interest as income on your return. This amount may represent interest on an overpayment for more than one year, or more than one kind of tax. This interest may have been paid with your tax refund or part or all may have been applied against other taxes you owed.

Exhibit 1

2021 Third Continental Congress
Office of the Secretary of State
Carpenters' Hall
Philadelphia, Pennsylvania
January 20, 2023

Greetings Coast Professional Inc.,

After March 4, 2023, the first order of business of the Third Continental Congress (TCC) is to assure each State National Citizen (SNC) of the 1776 Declaration of Independence and 1787 United States of America Constitution financial and economic retribution for the money Extorted from the SNC (a.k.a "American Tax Payers") since 1915 by the following Unconstitutional and illegal organizations and individuals (Defendants):

Violates Article I, Section 8, Clause 1 and/or Article XI of the 1787 United States of America Constitution.

1. Internal Revenue Service (IRS),
2. Bankrupted United States of America, INC. (USAINC),
3. Bankrupted United Nations, INC. (UNINC),
4. UNINC US Office of Personnel Management (OPM) (i.e. FBI, CIA, Congress, IRS, President, Supreme Court Justices, etc.)
5. British Accreditation Registry-Crown Temple British Maritain Law (B.A.R.)
6. The Central Bank of the USAINC is the Private Federal Reserve System, created by Congress in 1913.
7. Private Western Central Vatican Bank and Central Bank of London.
8. North Atlantic Treaty Organization (NATO)

All State National Citizens (a.k.a. American Tax Payers) shall not have any direct contact with any of the "Defendants" stated above. If one is approached by any individual claiming to representing the above stated "Defendants", do not be adversarial or aggressive, be polite and helpful. Report all encounters and all relevant correspondences (Postal Mail, E-Mail, Phone and personal contact reports) to the local Constitutional County Sheriffs' office.

All State National Citizens (a.k.a. American Tax Payers) shall:

1. Mail a copy your 2021 Tax Form (Personal or Business) originals to U.S. Department of the Treasury, National Payment Integrity and Resolution Center, P. O. Box 51315, Philadelphia, Pennsylvania, 19115-6314 and demand that all the money you have paid in taxes plus fees and interest (Since 1915, Example: Ratio for every $1 Tax paid times 10 = $10/$1) to the IRS be returned to you. Do not send any money.

2. Mail a copy of your tax forms, Blackout out your Signature and Social Security Number, to one of the following three (3) private collection agencies representing the IRS.

IRS Notice CP40 and Publication 4518: Effective September 23, 2021, when the IRS assigns your account to a private collection agency, one of these three agencies will contact you on the government's behalf:

   a. CBE Group Inc., P.O. Box 2217, Waterloo, IA 50704, 800-910-5837
   b. Coast Professional Inc., P.O. Box 526, Albion, NY 14411, 888-928-0510
   c. Conserve P.O. Box 307, Fairport, NY 14450, 844-853-4875

The Third Continental Congress Attorney General shall coordinate with the Governor of each of the 50 United States during the resolution of this Extortion Case and provide a progress report weekly.

All counter proposals from the Defendants, like Exhibit 1, shall be sent directly to: U. S. Department of the Treasury; National Payment Integrity and Resolution Center; P. O. Box 51315; Philadelphia, Pennsylvania 19115-6314.

Praise God for His gifts of Life, Liberty, and the Pursuit of Happiness to all mankind.

James N. Akins, Jr.
James N. Akins, Jr. © Without Prejudice, P. O. Box 1111, Fairhope, Alabama 36533-1111

2021 Third Continental Congress
Office of the Secretary of State
Carpenters' Hall
Philadelphia, Pennsylvania
January 20, 2023

Greetings Conserve,

After March 4, 2023, the first order of business of the Third Continental Congress (TCC) is to assure each State National Citizen (SNC) of the 1776 Declaration of Independence and 1787 United States of America Constitution financial and economic retribution for the money Extorted from the SNC (a.k.a "American Tax Payers") since 1915 by the following Unconstitutional and illegal organizations and individuals (Defendants):

Violates Article I, Section 8, Clause 1 and/or Article XI of the 1787 United States of America Constitution.

1. Internal Revenue Service (IRS),
2. Bankrupted United States of America, INC. (USAINC),
3. Bankrupted United Nations, INC. (UNINC),
4. UNINC US Office of Personnel Management (OPM) (i.e. FBI, CIA, Congress, IRS, President, Supreme Court Justices, etc.)
5. British Accreditation Registry-Crown Temple British Maritain Law (B.A.R.)
6. The Central Bank of the USAINC is the Private Federal Reserve System, created by Congress in 1913.
7. Private Western Central Vatican Bank and Central Bank of London.
8. North Atlantic Treaty Organization (NATO)

All State National Citizens (a.k.a. American Tax Payers) shall not have any direct contact with any of the "Defendants" stated above. If one is approached by any individual claiming to representing the above stated "Defendants", do not be adversarial or aggressive, be polite and helpful. Report all encounters and all relevant correspondences (Postal Mail, E-Mail, Phone and personal contact reports) to the local Constitutional County Sheriffs' office.

All State National Citizens (a.k.a. American Tax Payers) shall:

1. Mail a copy your 2021 Tax Form (Personal or Business) originals to U.S. Department of the Treasury, National Payment Integrity and Resolution Center, P. O. Box 51315, Philadelphia, Pennsylvania, 19115-6314 and demand that all the money you have paid in taxes plus fees and interest (Since 1915, Example: Ratio for every $1 Tax paid times 10 = $10/$1) to the IRS be returned to you. Do not send any money.

2. Mail a copy of your tax forms, Blackout out your Signature and Social Security Number, to one of the following three (3) private collection agencies representing the IRS.

IRS Notice CP40 and Publication 4518: Effective September 23, 2021, when the IRS assigns your account to a private collection agency, one of these three agencies will contact you on the government's behalf:

   a. CBE Group Inc., P.O. Box 2217, Waterloo, IA 50704, 800-910-5837
   b. Coast Professional Inc., P.O. Box 526, Albion, NY 14411, 888-928-0510
   c. Conserve, P.O. Box 307, Fairport, NY 14450, 844-853-4875

The Third Continental Congress Attorney General shall coordinate with the Governor of each of the 50 United States during the resolution of this Extortion Case and provide a progress report weekly.

All counter proposals from the Defendants, like Exhibit 1, shall be sent directly to: U. S. Department of the Treasury; National Payment Integrity and Resolution Center; P. O. Box 51315; Philadelphia, Pennsylvania 19115-6314.

Praise God for His gifts of Life, Liberty, and the Pursuit of Happiness to all mankind.

James N. Akins, Jr. © Without Prejudice, P. O. Box 1111, Fairhope, Alabama 36533-1111

2021 Third Continental Congress
Office of the Secretary of State
Carpenters' Hall
Philadelphia, Pennsylvania
January 20, 2023

Greetings CBE Group Inc,

After March 4, 2023, the first order of business of the Third Continental Congress (TCC) is to assure each State National Citizen (SNC) of the 1776 Declaration of Independence and 1787 United States of America Constitution financial and economic retribution for the money Extorted from the SNC (a.k.a "American Tax Payers") since 1915 by the following Unconstitutional and illegal organizations and individuals (Defendants):

Violates Article I, Section 8, Clause 1 and/or Article XI of the 1787 United States of America Constitution.

1. Internal Revenue Service (IRS),
2. Bankrupted United States of America, INC. (USAINC),
3. Bankrupted United Nations, INC. (UNINC),
4. UNINC US Office of Personnel Management (OPM) (i.e. FBI, CIA, Congress, IRS, President, Supreme Court Justices, etc.)
5. British Accreditation Registry-Crown Temple British Maritain Law (B.A.R.)
6. The Central Bank of the USAINC is the Private Federal Reserve System, created by Congress in 1913.
7. Private Western Central Vatican Bank and Central Bank of London.
8. North Atlantic Treaty Organization (NATO)

All State National Citizens (a.k.a. American Tax Payers) shall not have any direct contact with any of the "Defendants" stated above. If one is approached by any individual claiming to representing the above stated "Defendants", do not be adversarial or aggressive, be polite and helpful. Report all encounters and all relevant correspondences (Postal Mail, E-Mail, Phone and personal contact reports) to the local Constitutional County Sheriffs' office.

All State National Citizens (a.k.a. American Tax Payers) shall:

1. Mail a copy your 2021 Tax Form (Personal or Business) originals to U.S. Department of the Treasury, National Payment Integrity and Resolution Center, P. O. Box 51315, Philadelphia, Pennsylvania, 19115-6314 and demand that all the money you have paid in taxes plus fees and interest (Since 1915, Example: Ratio for every $1 Tax paid times 10 = $10/$1) to the IRS be returned to you. Do not send any money.

2. Mail a copy of your tax forms, Blackout out your Signature and Social Security Number, to one of the following three (3) private collection agencies representing the IRS.

IRS Notice CP40 and Publication 4518: Effective September 23, 2021, when the IRS assigns your account to a private collection agency, one of these three agencies will contact you on the government's behalf:

a. CBE Group Inc., P.O. Box 2217, Waterloo, IA 50704, 800-910-5837
b. Coast Professional Inc., P.O. Box 526, Albion, NY 14411, 888-928-0510
c. Conserve P.O. Box 307, Fairport, NY 14450, 844-853-4875

The Third Continental Congress Attorney General shall coordinate with the Governor of each of the 50 United States during the resolution of this Extortion Case and provide a progress report weekly.

All counter proposals from the Defendants, like Exhibit 1, shall be sent directly to: U. S. Department of the Treasury; National Payment Integrity and Resolution Center; P. O. Box 51315; Philadelphia, Pennsylvania 19115-6314.

Praise God for His gifts of Life, Liberty, and the Pursuit of Happiness to all mankind.

James N. Akins, Jr. © Without Prejudice, P. O. Box 1111, Fairhope, Alabama 36533-1111

2021 Third Continental Congress
Office of the Secretary of State
Carpenters' Hall
Philadelphia, Pennsylvania
January 20, 2023

Greetings Internal Revenue Service (IRS),

After March 4, 2023, the first order of business of the Third Continental Congress (TCC) is to assure each State National Citizen (SNC) of the 1776 Declaration of Independence and 1787 United States of America Constitution financial and economic retribution for the money Extorted from the SNC (a.k.a "American Tax Payers") since 1915 by the following Unconstitutional and illegal organizations and individuals (Defendants):

Violates Article I, Section 8, Clause 1 and/or Article XI of the 1787 United States of America Constitution.

1. Internal Revenue Service (IRS),
2. Bankrupted United States of America, INC. (USAINC),
3. Bankrupted United Nations, INC. (UNINC),
4. UNINC US Office of Personnel Management (OPM) (i.e. FBI, CIA, Congress, IRS, President, Supreme Court Justices, etc.)
5. British Accreditation Registry-Crown Temple British Maritain Law (B.A.R.)
6. The Central Bank of the USAINC is the Private Federal Reserve System, created by Congress in 1913.
7. Private Western Central Vatican Bank and Central Bank of London.
8. North Atlantic Treaty Organization (NATO)

All State National Citizens (a.k.a. American Tax Payers) shall not have any direct contact with any of the "Defendants" stated above. If one is approached by any individual claiming to representing the above stated "Defendants", do not be adversarial or aggressive, be polite and helpful. Report all encounters and all relevant correspondences (Postal Mail, E-Mail, Phone and personal contact reports) to the local Constitutional County Sheriffs' office.

All State National Citizens (a.k.a. American Tax Payers) shall:

1. Mail a copy your 2021 Tax Form (Personal or Business) originals to U.S. Department of the Treasury, National Payment Integrity and Resolution Center, P. O. Box 51315, Philadelphia, Pennsylvania, 19115-6314 and demand that all the money you have paid in taxes plus fees and interest (Since 1915, Example: Ratio for every $1 Tax paid times 10 = $10/$1) to the IRS be returned to you. Do not send any money.

2. Mail a copy of your tax forms, Blackout out your Signature and Social Security Number, to one of the following three (3) private collection agencies representing the IRS.

IRS Notice CP40 and Publication 4518: Effective September 23, 2021, when the IRS assigns your account to a private collection agency, one of these three agencies will contact you on the government's behalf:

a. CBE Group Inc., P.O. Box 2217, Waterloo, IA 50704, 800-910-5837
b. Coast Professional Inc., P.O. Box 526, Albion, NY 14411, 888-928-0510
c. Conserve, P.O. Box 307, Fairport, NY 14450, 844-853-4875

The Third Continental Congress Attorney General shall coordinate with the Governor of each of the 50 United States during the resolution of this Extortion Case and provide a progress report weekly.

All counter proposals from the Defendants, like Exhibit 1, shall be sent directly to: U. S. Department of the Treasury; National Payment Integrity and Resolution Center; P. O. Box 51315; Philadelphia, Pennsylvania 19115-6314.

Praise God for His gifts of Life, Liberty, and the Pursuit of Happiness to all mankind.

James N. Akins, Jr. © Without Prejudice, P. O. Box 1111, Fairhope, Alabama 36533-1111

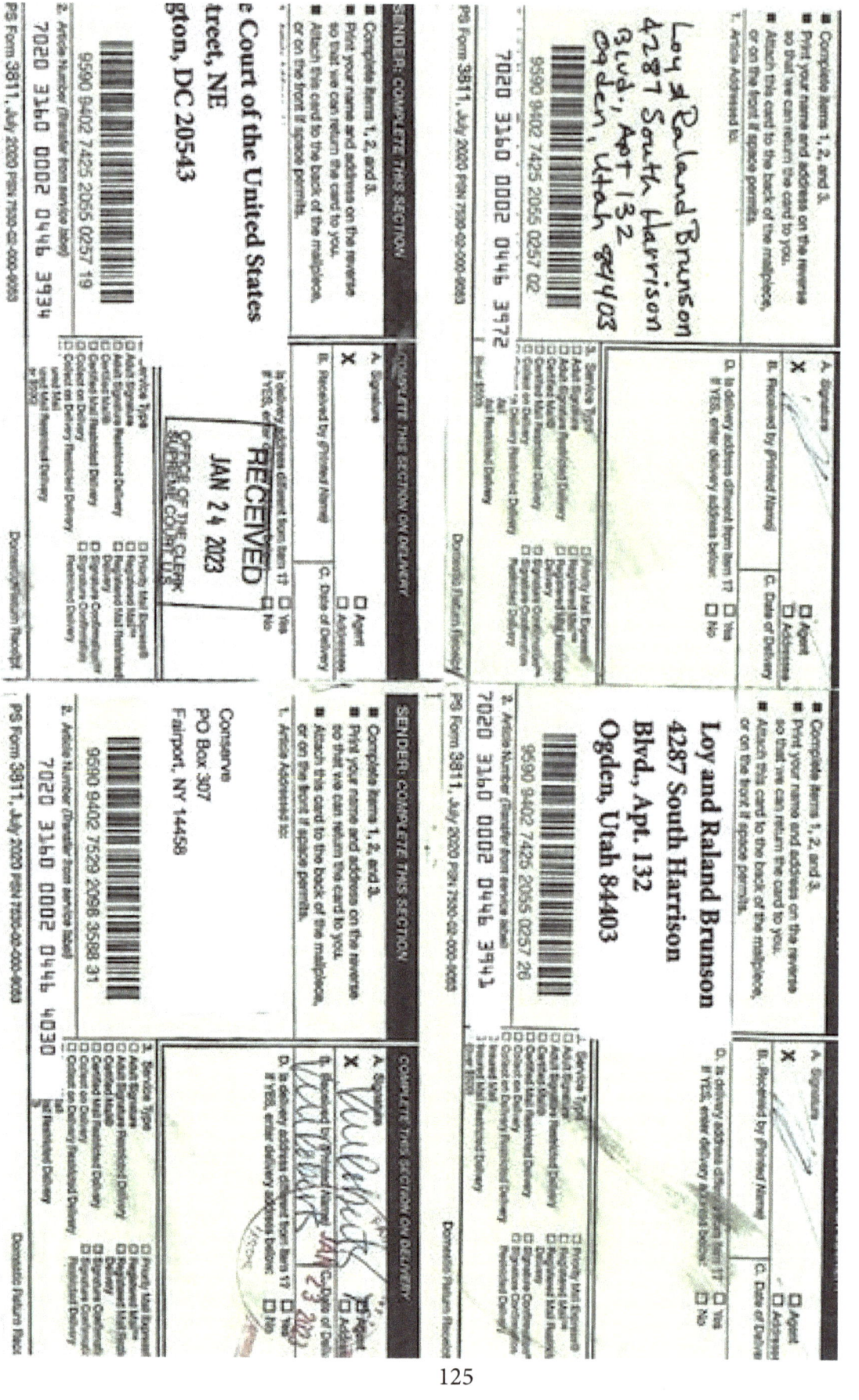

SENDER: COMPLETE THIS SECTION

- Complete items 1, 2, and 3.
- Print your name and address on the reverse so that we can return the card to you.
- Attach this card to the back of the mailpiece, or on the front if space permits.

1. Article Addressed to:

U. S. Department of the Treasury
National Payment Integrity and Resolution Center
P. O. Box 51315
Philadelphia, Pennsylvania 19115-6314

9590 9402 7529 2098 3588 00

2. Article Number *(Transfer from service label)*

7020 3160 0002 0446 4054

PS Form **3811**, July 2020 PSN 7530-02-000-9053

COMPLETE THIS SECTION ON DELIVERY

A. Signature
X _____ ☐ Agent ☐ Addressee

B. Received by *(Printed Name)*   C. Date of Delivery

COVID-19

D. Is delivery address different from item 1? ☐ Yes
   If YES, enter delivery address below: ☐ No

3. Service Type
☐ Adult Signature
☐ Adult Signature Restricted Delivery
☐ Certified Mail®
☐ Certified Mail Restricted Delivery
☐ Collect on Delivery
☐ Collect on Delivery Restricted Delivery
☐ Insured Mail
☐ Insured Mail Restricted Delivery (over $500)
☐ Priority Mail Express®
☐ Registered Mail™
☐ Registered Mail Restricted Delivery
☐ Signature Confirmation™
☐ Signature Confirmation Restricted Delivery

Domestic Return Receipt

Letter to the Truthers;

I know none of the "Truthers" will talk to me, because I am labeled a "Black Hat" fraud. But, that is just another way to separate us from knowing God's plan - Genesis 6:3, Daniel 2:35, and Book of Juan O'Savin, The Magador. In the last hours before Christ starts His 1000 years of peace, there are only God's Prophets, like Julie Green and me, Disciples of Christ, like Juan O'Savin and Satan and his dominions.

Like the Holy Bible, the book, "The Story of Our Life, Based on A True Life." is a history book not a story book. I have added proof that God wants His Disciples to tell His children of America, how to claim what has been stolen from them since 1915.

I went to www.piisthree.com and www.piis3.com, scrolled down to "Our Product", clicked on "FREE ebook", download Free ebook, Printed pages 114, 115, and 116, estimated how much money was stolen from my family since 1915 times 10 = $8,250,000.00, completed form documentation, follow mailing instructions in red on page 116.

I mailed my altered IRS 1040-SR to: U. S. Department of the Treasury, National Payment Integrity and Resolution Center and CBE Group, Inc. (IRS Private tax collection agent). I received a Form 1099-INT from Department of the Treasury, Internal Revenue Service, Memphis, TN 37501, for Calendar Year 2022, Total Interest Paid or Credited: $168.12.

I did not turn in a 2021 IRS 1040 or w2 form, therefore the IRS is responding to the altered IRS 1040-SR I submitted to National Payment Integrity and Resolution Center. Therefore, this is the IRS's resolution to my altered IRS 1040-SR. The address title of the IRS is Department of the Treasury, is not the same as U.S. Department of Treasury. The "Treasury" (i. e. Federal Reserve Bank, Central Banks) in the Department of the Treasury, is the Treasury of the United Nations, INC. Congress sign a 2018 bill to turn the US Office of Personnel Management over to the Secretary of State of the United Nations, INC.

When the Federal Reserve Bank's assets were put under the US Department of Treasury the United Nations, INC Treasury is EMPTY. Therefore, the debts of the Bankrupt USA, INC and Bankrupt DC INC have not been paid since October 1, 2022.

Since December 22, 2021, the US Constitutional Military, through the Third Continental Congress' 1787 US Constitution Commander in Chief President Donald J. Trump and the US Department of the Treasury, have been paying the US Office of Personnel Management employees and retirees.

All Disciples of Christ, please inform all We the People of The United States of America about the March 4, 2023 Swearing-in Ceremony of the 1st Session of the Third Continental Congress and how to claim the money that has been stolen from their families.

God has Blessed America, again.

2021 Third Continental Congress
Carpenters' Hall
Philadelphia, Pennsylvania
March 4, 2023

Greetings to all We The People of The Second Constitutional Republic of The United States of America:

By the Grace of God, all We the People of the 50 Continental United States of America are hereby invited to join the Second Constitutional Republic of The United States of America, in accordance with the 1787 US Constitution and 1776 Declaration of Independence with 2021 Article I and 2022 Article II. The Third Continental Congress (TCC), provisional 1787 United States Constitution government, was convened at Carpenters' Hall in Philadelphia, Pennsylvania on December 22, 2021.

On this date, March 4th, 2023 at 12:00 noon Central Standard Time, and the 72nd day of the 121st Jubilee of the 1000 year reign of our Lord Jesus Christ (Genesis 6:3 and Daniel 2:35), at the Swearing-in Ceremony for the 1st Session of the 2021 Third Continental Congress at the Shrine of the Lady Cross in Daphne, Alabama, on the Bay of the Holy Spirit:

We the People of the Second Constitutional Republic of The United States of America do hereby, in compliance with the 1787 United States Constitution, Article II, Section 2, Clause 1, inaugurate the Honorable Donald John Trump as the First President and Commander-in-Chief of the US Constitution Military.

The Honorable John C. White, Jr. of Arkansas has been sworn-in as 2021 Third Continental Congress (TCC) Speaker of the House of Representatives and shall be responsible for swearing-in the 200 TCC Congress members from all 50 United States of America.

According to tradition, I hereby resign my position of 2021 Third Continental Congress Secretary of State, to make way for President Donald John Trump to fill all his cabinet positions.

It has been my honor to have served God's children of The United States of America and praise God for the opportunity to do so. Since the Quantum technology in my book, "The Story of Our Life, Based on A True Life,," is from God, I will be transferring my registered Copyrights to the Quantum technology to the Second Constitutional Republic of The United States of America, where they may design, copyright, control worldwide operation and patents of Christ's Quantum Computers. Anyone in the world may download a FREE ebook by accessing www.piisthree.com or www.piis3.com, scrolling down to "Our Products" and click on "Free ebook"

Praise God for His gifts of Life, Liberty, and the Pursuit of Happiness to all Mankind.

James N. Akins, Jr. © Without Prejudice
2021 Third Continental Congress Secretary of State

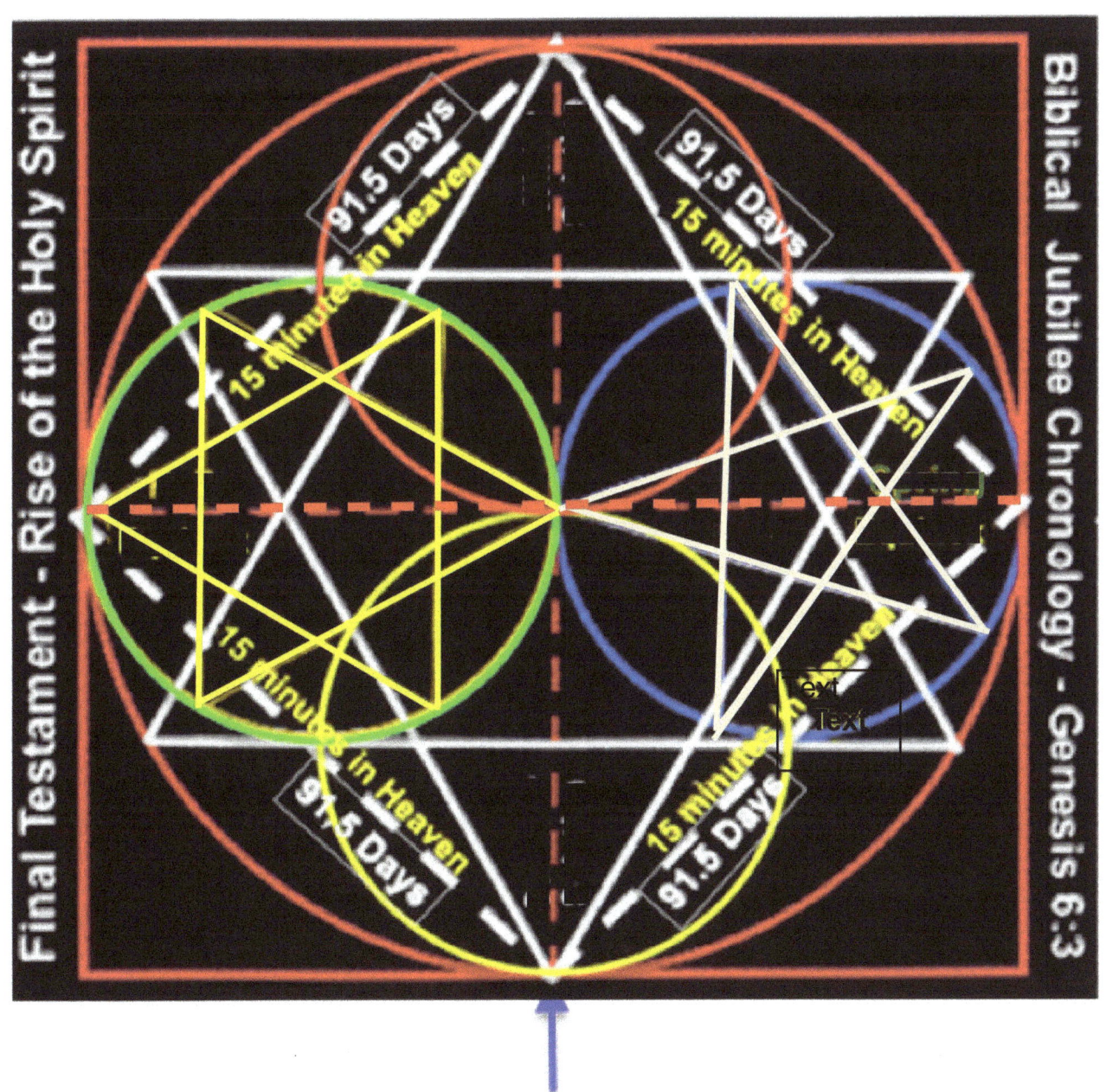

Final Testament – Rise of the Holy Spirit.

Our Lord Jesus Christ was born at the start of the 80th Jubilee December 22, 0000 just after midnight 00:00:00+. The end of the 120th Jubilee (6,022 years since God's Creation of Adam) (Genesis 6:3) was on December 21, 2022 at 24:00:00. The 2022th birthday of Our Lord Jesus Christ, on December 22, 2022 at 00:00.00+, marks the start of the 1st Positive (+) Jubilee of the Final Testament – Rise of the Holy Spirit.

I must apologize to Sir Isaac Newton for the uneducated statements I made in my book about him and his life's work. Sir Newton and I had three things in common; 1. We are just men of science that knew God is real and the Holy Bible is the written words of God. When the Holy Bible is decoded, it reveals the true Quantum mathematatica, astrology, Quantum Computers and software and physics of God's creations. Sir Isacc Newton sums up what I am saying in his two books: ****"Philosophiae Naturalis Principia Mathematica." In short, my perception of Newton's Calculus, as a lie or evil was wrong. Because, God wanted Satan and his dominions to think Newton's Calculus, as I point out in my book, was their means to rule over and destroy God's Creations through their creations; digital computers and Blockchain Quantum computers. They were wrong: God's Holy Bible WON!; 2. Both Sir Isaac Newton and I both came to the same conclusion, the Holy Bible recorded the True Advanced Technology that existed during the days of Moses and Noah. Sir Isaac Newton, Dr. Albert Einstein and Nikola Tesla were geniuses. However, most importantly they were profits in their days that 100% followed God's will through their Faith. Faith in God's will and Our Lord Jesus Christ is what we four guys have in common. and; 3. Our mothers do not understand us and God's plan for our life's contributions to Humanity. My mother thinks I am crazy.

**** © Newton, Isaac, Philosophiae Naturalis Principia Mathematica("Mathematical Principles of Natural Philosophy"), London, 1687; Cambridge, 1713; London, 1726. (Pirated versions of the 1713 edition were also published in Amsterdam in 1714 and 1723.)

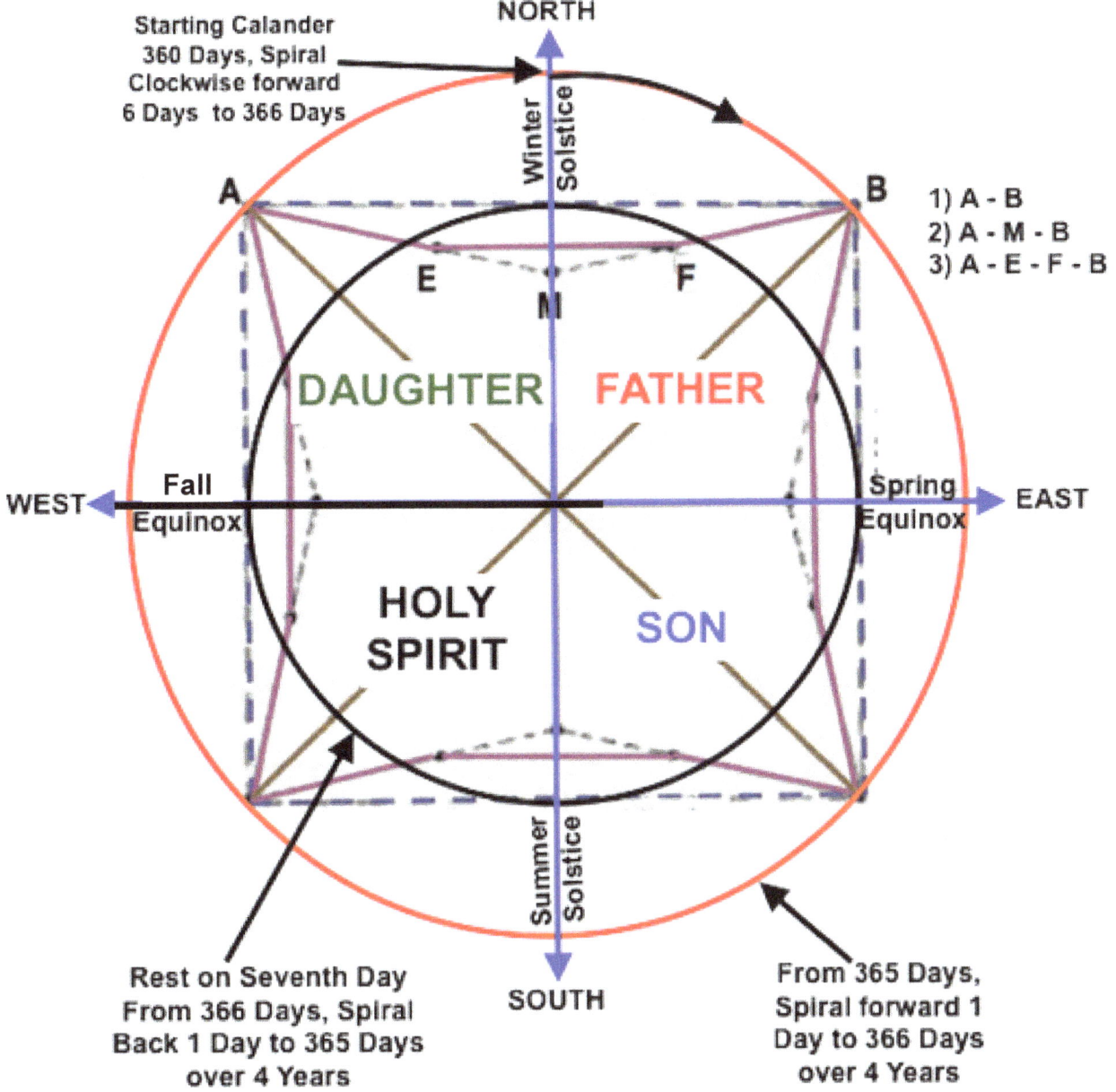

Pyramid of Khufu
Top Down View of Great Pyramid, Egypt